Synthesis Lectures on Mechanical Engineering

This series publishes short books in mechanical engineering (ME), the engineering branch that combines engineering, physics and mathematics principles with materials science to design, analyze, manufacture, and maintain mechanical systems. It involves the production and usage of heat and mechanical power for the design, production and operation of machines and tools. This series publishes within all areas of ME and follows the ASME technical division categories.

Steven F. Griffin · Daniel J. Inman

Design and Test of Dynamic Vibration Absorbers

Steven F. Griffin
The Boeing Company
Kihei, HI, USA

Daniel J. Inman
Department of Aerospace Engineering
University of Michigan
Ann Arbor, MI, USA

This work contains media enhancements, which are displayed with a "play" icon. Material in the print book can be viewed on a mobile device by downloading the Springer Nature "More Media" app available in the major app stores. The media enhancements in the online version of the work can be accessed directly by authorized users.

ISSN 2573-3168   ISSN 2573-3176 (electronic)
Synthesis Lectures on Mechanical Engineering
ISBN 978-3-031-43310-8   ISBN 978-3-031-43308-5 (eBook)
https://doi.org/10.1007/978-3-031-43308-5

© The Editor(s) (if applicable) and The Author(s), under exclusive license to Springer Nature Switzerland AG 2024

This work is subject to copyright. All rights are solely and exclusively licensed by the Publisher, whether the whole or part of the material is concerned, specifically the rights of translation, reprinting, reuse of illustrations, recitation, broadcasting, reproduction on microfilms or in any other physical way, and transmission or information storage and retrieval, electronic adaptation, computer software, or by similar or dissimilar methodology now known or hereafter developed.
The use of general descriptive names, registered names, trademarks, service marks, etc. in this publication does not imply, even in the absence of a specific statement, that such names are exempt from the relevant protective laws and regulations and therefore free for general use.
The publisher, the authors, and the editors are safe to assume that the advice and information in this book are believed to be true and accurate at the date of publication. Neither the publisher nor the authors or the editors give a warranty, expressed or implied, with respect to the material contained herein or for any errors or omissions that may have been made. The publisher remains neutral with regard to jurisdictional claims in published maps and institutional affiliations.

This Springer imprint is published by the registered company Springer Nature Switzerland AG
The registered company address is: Gewerbestrasse 11, 6330 Cham, Switzerland

Paper in this product is recyclable.

Springer Nature More Media App

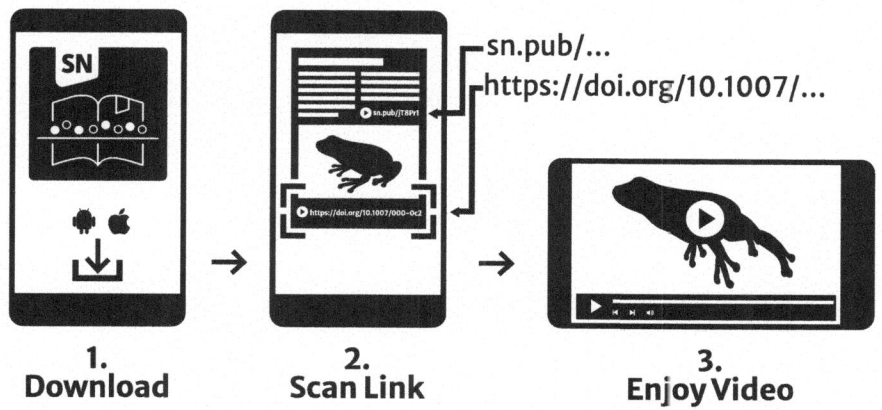

Support: customerservice@springernature.com

Contents

Design and Test of Dynamic Vibration Absorbers 1
1 Introduction 1
2 Description 3
   2.1   What Are Dynamic Vibration Absorbers and Why Are They Used? 3
   2.2   Construction of Devices 10
   2.3   Specification Worksheet 18
3 Analysis and Testing 19
   3.1   Analysis Methods: Equations of Motion 19
   3.2   Testing Methods 37
4 History and Lessons Learned 49
   4.1   Vertical Tails 49
   4.2   Self-tuning Mass Dampers 50
   4.3   Taipei 101 52
   4.4   MD-80/DC-9 Aft Cabin 53
   4.5   Passive VAs on Helicopters 55
5 Hands on Hardware Description and Instructions 56
   5.1   Host Structure 56
   5.2   Addition of a Tuned Mass Damper 56
   5.3   Addition of a Vibration Absorber 62
   5.4   Inexpensive Ways to Measure Vibration 65
References 68

Design and Test of Dynamic Vibration Absorbers

1 Introduction

All machines and structures vibrate and when their amplitude of vibration is too large, problems result varying from annoyance to fatigue, and eventually failure. For some devices, such as space telescopes and precision machines, large amplitude vibration can cause a severe lack of performance, rendering the system useless. Numerous methods can be used to mitigate large amplitude vibrations (see Table 1). Adding a damping material (usually viscoelastic) in a variety of different forms is well documented in the book, *Vibration Damping* [1]. A particularly effective treatment discussed is constrained layer damping, which consists of a stiff layer covering a viscoelastic material attached to a section of structure. The stiff outer layer causes more shear and hence more damping in the viscoelastic layer. Using active control methods is another way to suppress unwanted vibration amplitudes [2, 3]. However, material damping suffers from a lack of robustness to temperature changes, and active control requires an additional energy source and increases the complexity of hardware. Therefore, adding springs, masses and dampers is a useful, and often preferred solution. The focus here is on dynamic vibration absorbers (vibration absorbers and tuned mass dampers). Often a new system's vibration amplitudes are unknown at the design stage and turn up in prototyping or worse, manufacturing. Thus, an important class of vibration mitigation devices are added on after the original design. The addition of dynamic vibration absorbers to original systems to reduce vibration is the primary topic of this book.

Supplementary Information The online version contains supplementary material available at https://doi.org/10.1007/978-3-031-43308-5_1. The videos can be accessed individually by clicking the DOI link in the accompanying figure caption or by scanning this link with the SN More Media App.

Table 1 Possible ways to suppress forced vibration

Type	Comment
Vibration absorber	Added on after design, not in the load path
Tuned mass damper	Added on after design, not in the load path
Vibration isolator	Part of the original design, in the load path
Damping treatment	Added on after or during design, thickness constrained
Active control	Part of the original design, very effective but costly

Another successful means of reducing large amplitude vibration is called isolation. A vibration isolator is inserted in the load path to isolate the source of vibration from the system to be protected and is designed in terms of reducing either the force or displacement transmitted to the system of interest. It is important to note that isolation and absorption are very different phenomena.

Designing a dynamic vibration absorber consists of adding a lumped mass to the *host* structure, which is the structure or part in need of vibration suppression, and connecting it by a spring, effectively adding an additional degree of freedom to the host. There are two approaches to designing a dynamic vibration absorber system. The first is to connect the mass through only a spring, which is called a *Vibration Absorber* (VA). The second is to connect the mass through a spring and a damper, which we call a *Tuned Mass Damper* (TMD). Introductory vibration texts derive all the basic equations and models for these two approaches. The original idea is attributed to Frahm [4]. Both devices are often referred to as a *Dynamic Vibration Absorber* (DVA) [5].

A tuned mass damper consists of a mass connected to the host to be protected by damped springs and impedes local motion by increasing the damping of one or multiple vibration modes. In contrast, a vibration absorber consists of a mass connected to the host structure by a lightly-damped spring and impedes local motion by decreasing the displacement due to an input force at a single or very narrow band of input frequencies. The VA transforms the vibration of the host structure into oscillation of the absorber mass, essentially redistributing energy to decrease the host's response. On the other hand, a TMD takes energy out of the system and converts it to another form of energy through damping. The construction of both devices can be identical other than the necessity of a significant loss mechanism required for the design of the tuned mass damper. The mathematics involved in predicting the behavior of a host structure with an added vibration absorber or tuned mass damper are also identical other than the consideration of the loss mechanism in the tuned mass damper.

The VA and TMD have the benefit of lower risk and complexity compared with active control hardware. In many cases, tuned mass dampers and vibration absorbers can be used instead of active vibration control solutions saving costs.

2 Description

2.1 What Are Dynamic Vibration Absorbers and Why Are They Used?

Vibration and its effects are an inherent part of any structure or machine and are particularly prominent in vehicles. They can be both heard and felt at various levels throughout the vehicle. While some level of vibration is normal and expected, there may be a desire to reduce the level of vibration at certain locations to protect the structure or to improve human perception. DVAs are vibration mitigating devices used to limit the undesired vibration on all sorts of machines and structures where performance is reduced and/or fatigue becomes an issue if left unchecked.

Vibration Absorbers (VA) and Tuned Mass Dampers (TMD) have benefits for specific applications. Choosing which one to use can be done by considering the structure for which the dynamic vibration absorber is made, the dynamic response and the surroundings of the device.

More specifically, VAs should be considered when:

- The structural response is dominated by one or a small number of tonal disturbances that do not change in frequency.
- The direction of the response is constrained.
- There is room to install a relatively small device where it's desired that the structure would move less (often called a "rattle space").

Conversely, TMDs should be considered when:

- The structural response is dominated by one or a small number of lightly damped modes.
- The structure is very stiff and does not contain any "sweet spots" (regions of high in-plane strain), where constrained layer damping would be effective.
- Temperature extremes preclude viscoelastic materials because TMDs can be made of relatively temperature insensitive magnetic loss mechanisms and other materials robust to temperature gradients.
- There is room to install a relatively small device where the structure is moving the most (rattle space).

In general, TMDs are used to damp out lightly damped structural modes with broadband inputs and VAs are used to reduce the response to well defined disturbance tones (frequencies).

Key concepts in vibration reduction are that of frequency and of resonance. Every structure and machine has one or more natural frequencies [2]. For the simplest model consisting of a single mass of value m connected to ground through a spring of stiffness k, the *natural frequency* is defined as:

$$\omega_n = \sqrt{\frac{k}{m}} \tag{1}$$

in radians per second or

$$f_n = \frac{1}{2\pi}\sqrt{\frac{k}{m}} \tag{2}$$

in cycles per second or Hertz. If a harmonic force with a frequency of Ω and amplitude F is applied to the mass, the amplitude of the resulting vibration is

$$X = \frac{F/m}{\omega_n^2 - \Omega^2} \tag{3}$$

Note that the closer the excitation frequency gets to the natural frequency the larger this amplitude X becomes. When the two frequencies meet, Eq. (1) no longer holds, implying an increasing magnitude that is limited in real systems by a mechanical constraint, mechanical damping or by failure.

A good example of an aerospace application of a VA that is implemented on a heritage Boeing product is in the cabins of commercial aircraft, where cabin noise levels are high due to twin, aft engine vibration transmitted through the pylons to cabin skin structure, which radiates sound inside the cabin. Both frame and engine VAs can be used to reduce noise levels in these applications (see Sect. 4.4). In this case, the structural and acoustic response is dominated by a relatively constant frequency disturbance from the engine. The VAs do not damp the response but rather redirect the energy so that it is not transmitted into the host structure (the aircraft fuselage in this case). The energy shaking the host structure is redirected into shaking the absorber mass instead, resulting in the structure's vibration amplitude being reduced. Figure 1 shows a good example from Lang et al. [6] where the addition of absorbers reduced the acoustic response in the disturbance operating range but increased the response at frequencies where the engines did not excite the cabin. This example illustrates the trade necessary in a VA design and the importance of matching the device to the frequency of interest. In the time since this reference was published, more recent work has focused on active approaches to achieve the same or better objectives without the necessity of very accurate tuning [7].

An example of an aerospace application of a TMD that has been the subject of a great deal of research and testing is the vertical tail on fighter jets, where high angles of attack are a normal part of the flight envelope. High angles of attack result in unsteady flow due to vortex shedding at multiple or moving frequencies, which excites vibration modes in the vertical tails (see Sect. 4.1). This condition is encountered often enough to be responsible for fatigue damage at the base of the vertical tail. The frequency of the tail does not change very much with different flight conditions, and the introduction of a

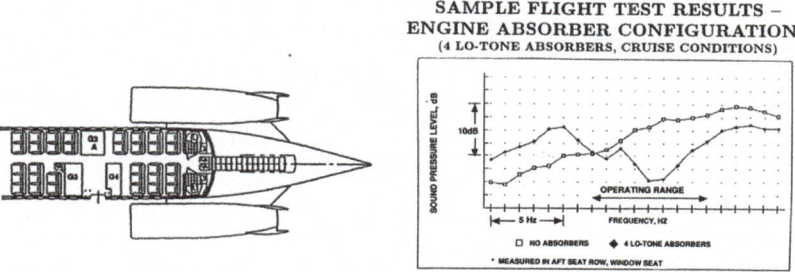

Fig. 1 Early MD 80s engine vibrations generated noise in the cabin. Absorbers were designed to cancel the vibration and hence mitigate the sound. The plot on the right shows sound pressure versus frequency with and without absorbers made during flight tests indicating the reduction of sound pressure in the operating range provided by the VA mounted on the engines

damper at a location of high displacement (the tip) greatly reduces the structural response of the relatively lightly damped tail without adding significant weight. In Fig. 2, Moog [8] illustrates the expected response on a vertical tail with and without a TMD.

If we plot increasing values of the excitation frequency, Ω, along a horizontal axis versus the amplitude normalized by the stiffness and the force (i.e., $X_1 k/F$) along the vertical axis, a plot like that of Fig. 2 results, called a *frequency response function*. This plot defines the concept of *resonance* as the maximum value of this curve, which occurs near $\omega_n = \Omega$.

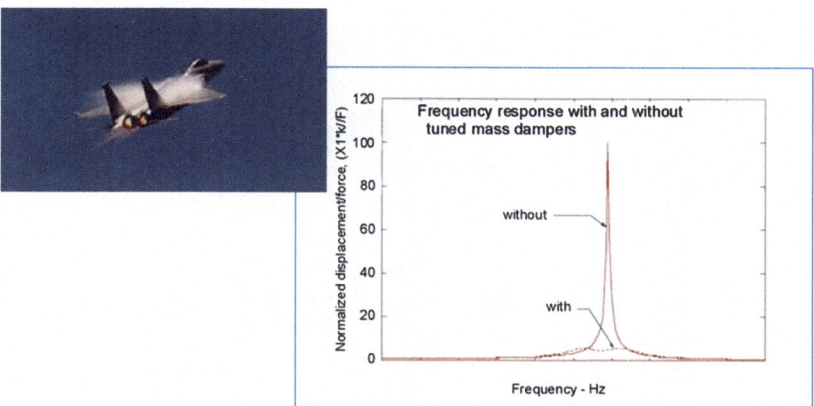

Fig. 2 The insert shows a twin tail fighter jet suffering from vibration due to fluid structure interaction. The plot on the right is the frequency response of a vertical tail with and without a TMD showing a clear reduction in vibration amplitude (Image courtesy of Moog)

The frequency response function plotted in Fig. 2 illustrates the large amplitude of vibration at the driving frequency (resonance) which is greatly reduced by the addition of the TMD. Note that one would expect that plotting Eq. (3) would result in an infinite value of X when the driving frequency is equal to the natural frequency. In fact, there is always some energy dissipation present and assuming a small amount of damping can be modeled as viscous, Eq. (3) becomes

$$X = \frac{F/m}{\sqrt{(\omega_n^2 - \Omega^2) + (2\zeta\omega_n\Omega)^2}}, \text{ where } \zeta = \frac{c}{2\sqrt{mk}} \quad (4)$$

Here ζ is called the damping ratio and the parameter c is called the viscous damping coefficient, usually a small approximate number based on measurements or in the case of a TMD, a designed amount of damping. The plot of Fig. 2 is formed from experimental data but also closely matches predictions made numerically by plotting $X_1 k/F$ versus Ω.

The first step in understanding how either a VA or a TMD works is to consider the basic spring mass damper system as described in any introductory vibration text and in the last chapter of any sophomore dynamics book. The schematic of a such a system is illustrated in Fig. 3.

If an alternating force of the form $F(t) = F\sin(\Omega t)$, is applied to m_1 in the vertical direction and its frequency is varied from close to 0 Hz to well above the resonance of m_1 on its spring-damper suspension, the resulting displacement, x_1, per applied force is shown in Fig. 4.

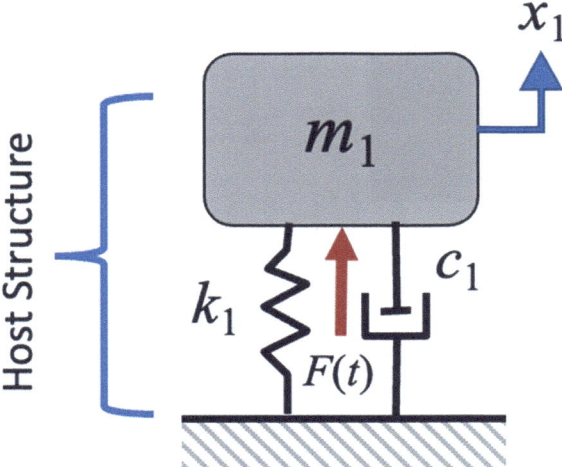

Fig. 3 Schematic the host structure modeled as a single degree of freedom system consisting of a mass, m_1, connected to ground by a spring k_1, and damper, c_1, subjected to a harmonic force

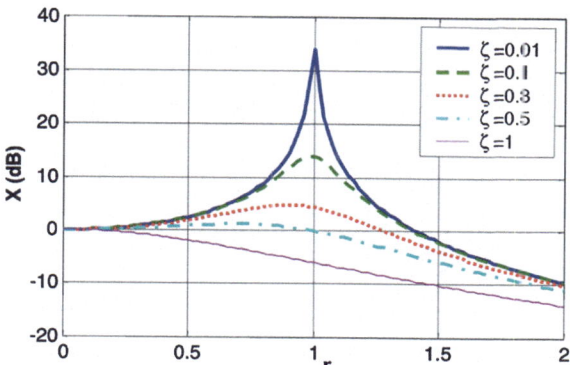

Fig. 4 The frequency response of the system in Fig. 3 subject to a harmonic force consisting of the magnitude versus the driving frequency divided by the natural frequency for various values of the damping ratio. As the damping increases the magnitude decreases

At low frequencies, the displacement is governed by the static stiffness of the spring. At resonance (normalized frequency $r = \Omega/\omega_n = 1$), the displacement is amplified by resonance and the maximum displacement depends on the damping in the system. In Fig. 4, the resonance on the left is lightly damped, and the displacement is amplified by a factor of almost 40 over its static response.

If an unwanted disturbance was applied to m_1 in the x_1 displacement direction that had frequency content at or near the resonant frequency, most of the displacement would be at or near the resonant frequency, and the displacement amplification may be unacceptable. This is the case for the vertical tail shown in Fig. 2, where the tail is the spring mass system, and the disturbance is the flow created at high angles of attack. If a much smaller spring mass damper is added to the spring mass damper in Fig. 3, the schematic in Fig. 5 results, leading to the frequency response plot given in Fig. 6. The mass labeled m_1 with its corresponding spring and damper represents a crude approximation of the structure in need of vibration mitigation. This is sometimes called the *host structure* or the *primary structure*. The mass m_2 and its connecting spring and damper are called the absorber system and it is designed to mitigate the vibration of the primary structure. Figure 5 also illustrates the difference between the VA and TMD. In the VA the damper connection to the host structure (m_1) is largely ignored. In the TMD design, the damping coefficient c_2 is a major design parameter.

The absorber system can be designed to have a much higher damping but approximately the same frequency as the primary mass taken by itself (see Sect. 3). The result is that the original lightly damped resonance of the primary structure couples with the heavily damped resonance of the absorber mass. A new system results in two moderately damped, coupled resonances as shown in Fig. 6 and compared to the original response from Fig. 4. The design of the absorber mass spring damper to couple with the primary

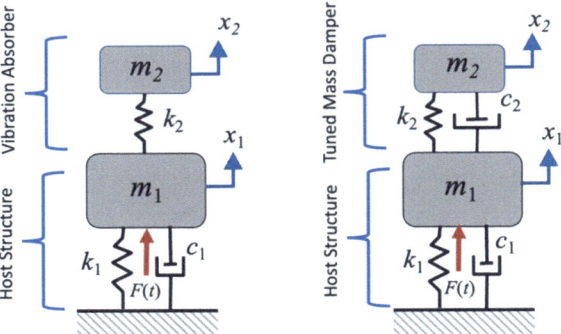

Fig. 5 Two schematics illustrating the difference between adding a VA and a TMD to the structure model of Fig. 3 consisting of a large mass, m_1, connected to ground through a spring k_1 and damper c_1. The VA is illustrated on the left and the TMD is illustrated on the right

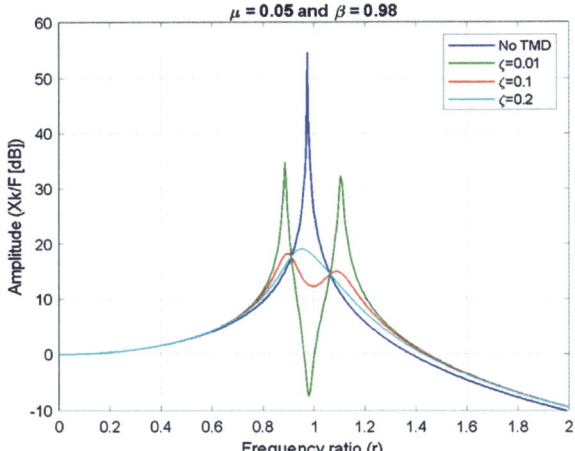

Fig. 6 The amplitude response of the primary mass, m_1, versus the frequency ratio without a TMD (blue line) along with the response with a TMD for increasing values of damping ratio for given a fixed mass ratio, $\mu = m_2/m_1$, and frequency ratio $\beta = \omega_2/\omega_1$

mass by adding damping is called a TMD. The right side of Fig. 5 illustrates a model for a TMD. The term "tuned" in TMD refers to tuning the vibration frequency of the TMD so that it most efficiently adds damping to the vibration frequency of the primary structure. The amount of added mass, the uncoupled damping in the TMD, and the uncoupled resonance of the TMD dictate the amount of reduction of resonant amplification that can be achieved.

In discussing various absorber designs it is useful to define the mass ratio and frequency ratio of the two masses. The mass ratio, μ, is defined as the ratio of the absorber mass to the primary mass and is given by

2 Description

$$\mu = \frac{m_2}{m_1}$$

In mass restricted applications the ratio μ is restricted to be small. The other useful ratio is the decoupled frequency ratio, β, defined as the ratio of the natural frequency of the absorber to the natural frequency of primary mass, considering each as a single degree of freedom system by

$$\beta = \frac{\omega_2}{\omega_1} = \frac{\sqrt{k_2/m_2}}{\sqrt{k_1/m_1}}$$

Using the two constants, μ and β, and realizing the static deflection of the primary mass is $X_s = F/k_1$, the ratio of the amplitude of the primary mass, X_1, to the static deflection can be written completely in terms of dimensionless quantities as

$$\frac{X_1}{X_s} = \sqrt{\frac{(2\zeta r)^2 + (r^2 - \beta^2)^2}{(2\zeta r)^2(r^2 - 1 + \mu r^2)^2 + [\mu r^2\beta^2 - (r^2 - 1)(r^2 - \beta^2)]^2}} \qquad (5)$$

If, on the other hand, the disturbance was applied well below resonance, at one half the resonant frequency shown in Fig. 4, the resulting displacement would only be slightly larger than the static displacement. The frequency and amplitude of the disturbance can cause unwanted sound radiation. This is exemplified by the MD 80 from Fig. 1. In this case, the disturbance is due to some slight motor imbalance that is located very close to the aircraft skin/trim panel. This causes the skin to vibrate and radiate sound to the passengers sitting nearby. The small spring mass damper can be designed so that its uncoupled resonance, ω_2, occurs at a frequency very close to the disturbance frequency. The result is a decrease in the response at the frequency of the uncoupled resonance. The amplitude of the coupled response at the disturbance frequency decreases with less damping in the VA. Unlike a TMD, a VA is usually designed as a "high Q" or lightly damped spring-mass (Q = $1/2\zeta$). Like the TMD, the amount of mass added also establishes a limit on the amount of reduction that can be achieved. The resulting dip in response at the frequency of the disturbance is shown with the original response in Fig. 7.

In the case of the TMD, a reduction in the resonant response occurs due to added damping in the absorber design. Because this treatment removes energy from the system analogous to a resistor in an electrical circuit, the TMD is sometimes referred to as a resistive device. In the case of the VA, we have a reduction in the response away from resonance too due to the effect of a lightly damped resonance creating a reduction at a specific frequency, called a zero in control theory. Because this treatment does not remove energy but redirects it, analogous to capacitors and inductors in an electrical circuit, the VA is sometimes referred to as a reactive device. The design and construction of the TMD and the VA might be identical other than the inclusion of a relatively large loss mechanism in the design of a TMD and the minimization of loss mechanisms in the design of a VA.

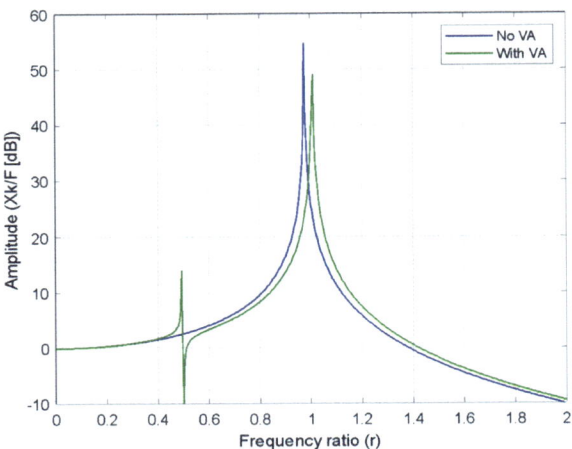

Fig. 7 The result of adding a VA tuned to one half the resonance frequency is to create a large reduction in response at the disturbance frequency, cause a slight shift and lowering of the peak amplitude and to introduce a second peak slightly below the disturbance frequency

2.2 Construction of Devices

Both TMDs and VAs must have a mass and spring component. The TMD also requires a significant loss mechanism component. Aside from the loss mechanism, the construction of the devices may be identical. Both devices may be constructed to address linear or angular motion, which would then dictate the construction of the spring, mass and damper components. In the case of an angular damper or absorber, the inertia of the added device determines the maximum reduction as opposed to the mass in a linear device. Since most mechanical and structural applications of dynamic vibration absorbers are concerned with reduction of linear motion, absorbers of linear motion form the focus of this book. However, many of the same principles can be used to suppress angular motion.

An added benefit of both TMDs and VAs is that they do not change the load path between the disturbance and the location where vibration is reduced by adding the device. An example of this where a TMD might be considered is a precision device like an airborne telescope that has a mirror vibrating on a lightly damped suspension. To stop the mirror from vibrating, it may be possible to stiffen the suspension, but this would likely impact the optical performance of the mirror, assuming it was optimized to minimize optical wavefront distortion. It may also be possible to insert a loss mechanism between the mirror and the mount, but this would limit the ability to hold a precise static position. If a TMD were added to the back of the mirror to suppress its motion, it would add damping to the resonance without causing distortion or adding uncertainty to its position. This mitigation is unique in not forcing a trade between optical performance and vibration performance. An example where a VA might be considered is a vertical lift vehicle with an identified periodic input that is caused by a harmonic of the rotor turning frequency. If there is excessive vibration at a location on the vehicle, it could potentially be reduced by adding stiffness or incorporating mechanical isolation. Either change would cause a change in the load path of the vehicle that would force consideration of whether stress criteria or other design criteria are still satisfied. However, the addition of a VA has the potential to not change the load path and still reduce vibration.

A low frequency limitation must be considered when contemplating the use of either a TMD or a VA to fix a vibration problem when the direction of interest has a component of gravity acting on the inertial mass, m. If we consider the case of a TMD or VA oriented vertically, its inertial mass will be acted on by the force of gravity, f, by

$$f = mg \tag{6}$$

where g is the acceleration due to gravity. The spring force resisting gravity is

$$f = kx \tag{7}$$

where k is the spring constant of the TMD or VA and x is the static deflection. If we consider the primary system of Fig. 3, decoupled from the absorber, k can be further represented in terms of the natural frequency and mass as

$$k = (2\pi f)^2 m \tag{8}$$

Taking Eqs. 6–8 and solving for x as a function of f gives

$$x = \frac{g}{(2\pi f)^2} \tag{9}$$

Equation 9 shows the vertical, static deflection of a mass due to gravity can be completely specified by the vertical bounce frequency (frequency at which the mass will oscillate if the spring is compressed). The lowest frequency these devices can be designed for is usually in the neighborhood of 10 Hz, where, according to Eq. 9, static deflection is

$$\frac{386 \frac{\text{in}}{\text{s}^2}}{(2\pi 10)^2} = 0.1'' \tag{10}$$

Designing a TMD or VA to accommodate more than 0.1" of static sag becomes problematic for many reasons, including required space and nonlinearity in the spring. Dynamic deflections may also be quite large at low frequencies which may impose another low frequency limit. If gravity acts in a direction orthogonal to the motion of the device, a much lower frequency problem can be solved with a TMD or VA. This is usually the case for civil structures that incorporate TMDs. A semi-active device has been developed that uses an active servo loop to keep the mass centered in the presence of gravity [9]. This approach works only with TMDs that incorporate a voice coil/magnet as a loss mechanism and requires some additional hardware and power but can achieve extremely low resonant frequencies in the vertical direction in the presence of gravity.

2.2.1 Springs

All TMDs and VAs are connected to the primary mass by a spring. The choice of spring depends on the amplitude of motion and the space available for the absorber mass to oscillate in. Springs can be classified into two major types, either a flexure or a mechanism. A flexure is a spring that does not have any moving parts and motion results from bending of the part. A mechanism has moving parts and usually has some interface where there is some amount of friction. A flexure is generally good for small motions, whereas a mechanism can be good for large motions. Figure 8 shows two examples of flexures and mechanisms that might be used for linear TMDs or VAs.

Flexures

When designing a flexure, it is important to make sure it does not plasticly deform and will stay in its linear range and hence function properly. To maximize the linear range of a flexure, shaping can minimize stiffening that occurs with deflection and maximize its usable range, making it good for moderate deflections. Taking the example of a 0.02″ thick, steel flexure with the spring action designed for motion normal to the page (out-of-plane motion), two examples are shown with the same out-of-plane stiffness and same boundaries on the outside as a constraint and on the inside as a point of uniform application of force in Fig. 9. The out-of-plane, linear motion that these flexures are designed to accommodate will be referred to as axial motion.

The blue flexure in Fig. 9 has been shaped to reduce stress for the same axial deflection as the orange flexure. A linear static stress analysis of both the flexures in Fig. 9 is shown in Fig. 10 for the same applied load.

The maximum stresses shown in Fig. 10a illustrate that the axial flexure on the left-hand side experiences a much higher stress for the same load compared to the shaped flexure in Fig. 10b. A nonlinear analysis shows a similar stress increase for the straight

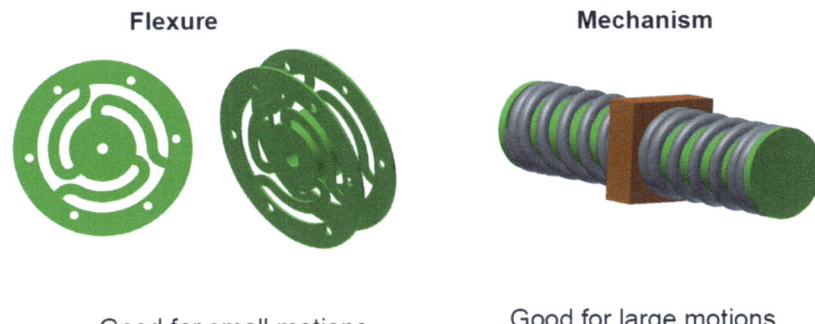

Fig. 8 Examples of two types of springs used for TMDs and VAs used to suppress axial motion. The flexure style illustrated on the left works well for relatively small displacements, whereas mechanisms (right) are needed to handle larger deflections

Fig. 9 Two different kinds of flexures, straight (orange) and shaped (blue). Flexures constitute the spring k_2 and are used to mount the absorber mass, m_2, on the center point. The different shape of the flexures allows versatility to address stress and fatigue issues

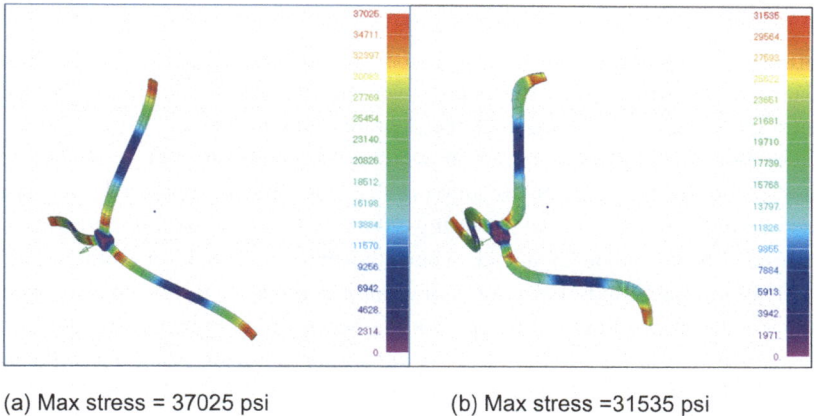

(a) Max stress = 37025 psi   (b) Max stress =31535 psi

Fig. 10 The stress distribution on the two types of flexures for same load. **a** the axial flexure and **b** the shaped flexure. The stresses help determine the type of material to use and the best flexure shape

flexure on the left but also shows the center rotating slightly about the z axis with axial displacement for the curved flexure on the right. Since both the tuned mass damper and the VA are not affected by a slight rotation of their suspended mass, this does not detract from device performance. In a dynamic device the higher stress in straight flexure couples stress induced by axial motion to in-plane stiffening, making a very nonlinear stiffness that is dependent on amplitude.

Another requirement on most flexure devices is to share the bending stiffness with two flexures as is shown in the flexure example of Fig. 8. This constrains the suspended mass to axial motion and constrains tilt in the x and y axes. This is usually necessary to drive vibration modes of the device with motion other than axial to a high enough frequency so that they do not have to be considered in the design. The design of flexures that satisfy all these requirements usually requires a finite element model.

Because the flexures in dynamic vibration absorbers are part of a device that controls vibration, not only strength but also fatigue must be considered in their design. This usually drives the designer toward high fatigue strength materials. Common examples are steel optimized for strength and fatigue like Aermet [10], or more exotic materials like titanium when strength-to-weight ratio is also important.

Another feature that can be incorporated into a flexure is a means of changing the frequency of the device. A simple way to do this is to provide multiple thickness copies of the same flexures that can be used interchangeably when tuning the device. Commercial devices are available that incorporate a variable length flexure that can be changed once the device is installed when tuning the device [11] and locked into a final configuration once the desired behavior is achieved. Adding or subtracting mass is also an effective means of changing the frequency of the device, especially if small changes are needed.

Mechanisms

If a flexure cannot be designed to operate over the required stroke for a TMD or VA, it is necessary to design a spring that incorporates moving parts. One example of this is shown in Fig. 8. The helical spring shown can accommodate a large stroke, but it is necessary to also include a means of moving the mass that is supported by bearings to minimize sliding friction. Even with bearings, there will always be a nonlinearity at very low amplitudes in this style of device where the mass is "stuck" at some equilibrium position. This low amplitude behavior will exhibit a mode at low frequency where the mass moves without breaking the friction on the bearings. At higher amplitudes, the low frequency mode will be replaced with a classic spring behavior. Since these devices are designed for large amplitude behavior, this behavior may not impact overall performance, but a dynamic vibration absorber that incorporates a mechanism should not be expected to work well at low amplitudes.

2.2.2 Loss Mechanisms

Unlike VAs, a critical component of every TMD is the loss mechanism, i.e., the damper in Fig. 5. By far the most common loss mechanism in consumer products for relatively small motion is viscoelastic damping that results from including elastomeric materials in the design. Elastomers are rubber or rubbery polymers that display viscoelastic damping behavior when they are deformed. This deformation, even at very small amplitudes, results in internal friction due to relative motion of long, polymeric cross-linked chains. This friction is converted into heat and the damping can be viewed as a conversion of vibration energy to heat energy. Common rubbery polymers like sorbothane or neoprene are often incorporated into grommets to serve as both a spring and a damper to isolate delicate equipment. For a TMD, viscoelastic damping can be incorporated by adding constrained layer damping to the flexures [1]. As with grommets, there are many damping materials available from companies like 3M for use in constrained layer damping. The use of viscoelastic damping as the loss mechanism in a TMD is simple and well understood. The problem is that the damping behavior of a polymer is extremely temperature dependent, and many applications require that the TMD operate over a broad temperature range. In that case, it is necessary to consider other loss mechanisms.

Another loss mechanism that may be employed is fluid viscous friction. This is accomplished by moving a fluid through passages and/or valves within the TMD or a discrete component of the TMD, usually referred to as a damper. The fluid motion is usually tied to the motion of a piston that, in turn, is coupled to the relative motion of the moving mass to its housing. Although discrete dampers in aerospace applications are not common, they are used in TMDs that damp building vibrations. An example of this is the primary Taipei 101 tuned mass damper shown in Fig. 11, where the large mass is suspended as a pendulum and discrete dampers connect the mass to the surrounding structure.

Since the damping force generated is proportional to rate for laminar fluid flow, viscous friction is the ideal loss mechanism. Like elastomers, nearly all working fluids have damping performance that depends on temperature. In the case of a liquid, the most important parameters related to damping are density and viscosity. In the case of gases, like air or

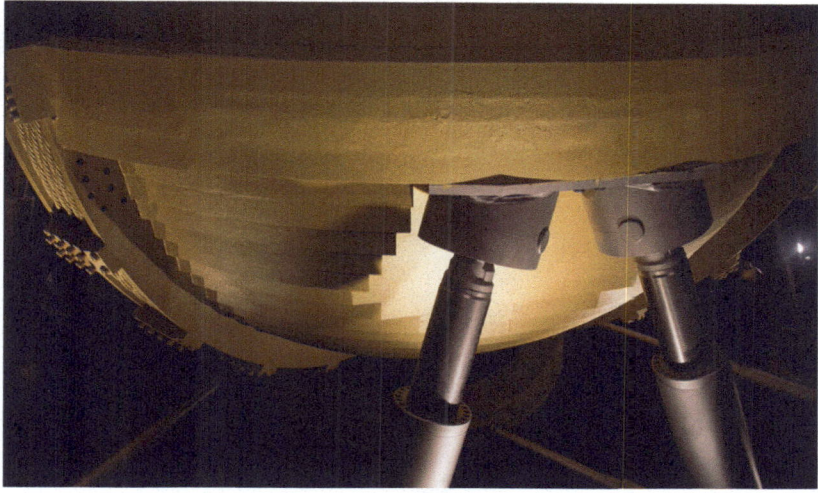

Fig. 11 Taipei 101's tuned mass damper. The gold ball is the mass m_2 which is suspended to swing like a pendulum connected to dampers (the grey rods)

nitrogen, compressibility and density are important parameters in dashpot design that performs the function of a damper. Although these may be less sensitive to temperature, in aerospace applications where the dashpot draws air in and out from the ambient surrounding air, altitude may become an important factor on the damping that may be achieved. Commercial vendors like Airpot [12] sell a wide range of discrete air dampers. When the damper is intended to perform in a pressurized, ambient environment, air dampers have the advantage of never leaking, since it uses ambient air as its working fluid. This limits aerospace applications where the ambient air pressure and density might change dramatically with altitude. Another important consideration on commercial air dampers is that they may include linkages with free play as part of their interface. In some cases, this translates to a lower amplitude limit where the damping mechanism is not effective. An innovative TMD that uses gas viscous friction as the loss mechanism for wind tunnel testing, where ambient temperature does not vary, uses a tunable spring made of a shape memory alloy to allow for changes in the frequency of the tuning mass with the application of a temperature change to the spring [13].

A loss mechanism that is practically independent of temperature over most applications' operating range is magnetic damping. Magnetic damping also has the advantage of having very linear viscous damping characteristics. A magnetic damper uses the interaction between the motion of a magnet and a conductive material like copper or aluminum. A common example that illustrates this concept is the voice coil motor shown in Fig. 12.

When a current passes through the coil of wire shown in Fig. 12, a magnetic field develops that is either in the same or the opposite direction of the permanently poled magnet in the middle of the device. This interaction between magnetic fields causes the magnet and/or coil move in the same direction as the induced magnetic field when the directions agree or the opposite direction when the directions are opposed. In this way, the application of current can induce relative motion between the magnet and the coil.

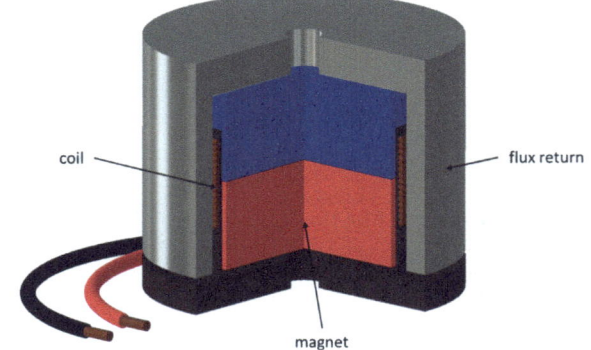

Fig. 12 An alternate way to create a damper is to use to allow a magnetic mass to oscillate through a coil of wire, or a voice coil motor. The magnetic field creates a electric field in the coil retarding the motion proportional to its velocity

For magnetic damping, the concept is reversed. If we move the magnet in the vertical direction in Fig. 12 relative to the coil and provide an electrical connection between the wires, current flows through the resulting circuit with the direction of the current being determined by the direction of the motion of the magnet. Since voice coil motors always have an inherent resistance, the maximum current flow corresponds to shorting the wires and moving the magnet. If a resistor is placed between the two wires of the voice coil, the amount of resistance will not only determine the amount of current flow but also the amount of magnetic damping. This is because current will be dissipated in both the inherent resistance of the coil and the added resistor and converted to heat. The equation that gives the damping coefficient that results from using a voice coil in reverse is

$$c = \frac{K_f K_b}{R} \qquad (11)$$

where K_f is force sensitivity, K_b is the back EMF constant and R is the sum of the coil resistance and any added resistance in series with the coil. The resulting units are force per velocity. If the added resistance is a potentiometer, a means is provided of changing the damping the magnetic loss mechanism provides by adjusting the resistance.

Similar to the viscoelastic loss mechanism, the magnetic damper works by converting mechanical motion into heat. Since the "correct" damping of a TMD is often determined as part of the tuning process, making the added resistance variable is helpful during initial tuning. If the coil is replaced with a conducting cylinder, and the magnet is moved with respect to it, magnetic damping also results but without the ability to conveniently change the value during tuning. In this case, the damping results due to the generation of eddy currents within the conductive material. This can be looked at as current flow at a micro level in the case of the conductive cylinder as compared to current flow at a macro level in the case of the coil. If high damping levels are required for the TMD, the solid cylinder is preferable to the coil due to the ease of incorporating a higher density of conductive material in close proximity to the magnet. In either the coil or the solid cylinder case, magnetic damping degrades at high frequencies due to the frequency dependence of the electrical impedance in the electromagnetic circuit. For instance, Moog shows a working range up to 120 Hz on its Commercial Off The Shelf (COTS) devices [11] with magnetic loss mechanisms. A patented approach [14] to achieve higher damping for the same magnet density and a higher working frequency range that involves a unique arrangement of the magnets relative to a solid conductive cylinder. This approach illustrates that when the goal is damping, as opposed to an architecture that was based on transduction, the magnetic field can be tailored to achieve more damping for the same weight.

Loss mechanisms to achieve damping in TMDs are summarized in Table 2 along with their advantages and disadvantages.

Table 2 Loss mechanism comparison

	Advantages	Disadvantages
Elastomer	• Inexpensive • Simple to incorporate • COTS • Can work as spring and damper	• Temperature dependent • Potential for outgassing • Durability
Liquid fluid	• COTS as discrete damper (Civil structures)	• Temperature dependent • Hard to incorporate into device • Potential for leaks
Gas fluid	• COTS as discrete damper	• Temperature dependent
Magnetic	• Little or no temperature dependence	• Decreasing effectiveness at high frequencies

2.3 Specification Worksheet

When procuring or constructing a TMD or VA for a given application, the worksheet in Fig. 13 gives a minimum set of specifications to consider.

```
Mechanical Requirements
Stiffness: k=_____ lb/in    Tolerance: ± _____ lb/in
Damping: k=_____ lb/(in/s)  Tolerance: ± _____ lb/(in/s)
Suspended mass: m=_____ lb mass  Tolerance: ± _____ lb mass
Uncoupled primary (first) resonance: f=_____ Hz  Tolerance: ± _____ Hz
Uncoupled second resonance: f>_____ Hz  Tolerance: ± _____ Hz
Linear range
Total range and linearity requirements (provide curve)
Static deflection in preferred orientation
Stiction range if any
Fatigue considerations

Installation interface and volume requirements (provide solid model)

Environmental requirements
Temperature
Altitude
Sand and dust
Humidity
Salt Spray
Hydraulic fluids
Magnetic field if any
```

Fig. 13 A typical specification worksheet to start the design of a TMD or VA based on minimum details. Note that the units should be changed for use in other countries and be careful not to confuse radians and Hertz when making frequency calculations and measurements

3 Analysis and Testing

3.1 Analysis Methods: Equations of Motion

The first step to analyzing the effect of a VA or TMD on a structure is to assemble the equations that represent the model in Fig. 5. Summing forces and solving the resulting set of equations yields the following equations of motion

$$\begin{bmatrix} F \\ 0 \end{bmatrix} = \begin{bmatrix} m_1 & 0 \\ 0 & m_2 \end{bmatrix} \begin{Bmatrix} \ddot{x}_1 \\ \ddot{x}_2 \end{Bmatrix} + \begin{bmatrix} c_1 + c_2 & -c_2 \\ -c_2 & c_2 \end{bmatrix} \begin{Bmatrix} \dot{x}_1 \\ \dot{x}_2 \end{Bmatrix} + \begin{bmatrix} k_1 + k_2 & -k_2 \\ -k_2 & k_2 \end{bmatrix} \begin{Bmatrix} x_1 \\ x_2 \end{Bmatrix} \quad (12)$$

From this, the relationship for the amplitude of the motion of the primary mass, X_1, given the force magnitude F can be found to be:

$$\frac{X_1}{F} = \sqrt{\frac{(k_2 - m_2\omega^2)^2 + \omega^2 c_2^2}{[(k_1 - m_1\omega^2)(k_2 - m_2\omega^2) - m_2 k_2 \omega^2]^2 + (k_1 - (m_1 + m_2)\omega^2)^2 \omega^2 c_2^2}} \quad (13)$$

where m_1 and m_2 are the primary and absorber masses respectively, c_2 is the damping coefficient of the damper attached to the absorber mass and k_1 and k_2 are the spring constants of the springs attached to the primary and absorber masses respectively. The force $F(t)$ of magnitude F is applied to the primary mass and the introduction of the absorber mass is intended to change the response of the host mass to the applied force. Assuming the damping associated with the primary mass, c_1, is relatively small and results in a lightly damped mode of the primary mass without the introduction of an absorber mass, we have two possible cases representing dynamic vibration absorbers:

- TMD—c_2 is relatively large and m_2 is between 1/10th and 1/100th the mass of m_1 and the uncoupled natural frequency of the primary mass is at or slightly below that of the large mass (i.e., $0.1 \geq \mu \geq 0.01$)—Fig. 6.
- VA—c_2 is very small and m_2 is between 1/10th and 1/100th the mass of m_1 (i.e., $0.1 \geq \mu \geq 0.01$) and the uncoupled natural frequency of the absorber mass is very different from that of the large mass (β is relatively small, $\beta < 1$) as in Fig. 6.

If all of the applications of interest could be abstracted as a single, host mass with a collocated disturbance, the analysis of a solution could be completed using the simple Eq. 13. The values of k_2, c_2 and m_2 could be varied until the desired response is achieved as reflected by the resulting frequency response function. If the disturbance is well known, the expected reduction of the disturbance due to adding the DVA could be predicted in both the time or frequency domain.

3.1.1 Frequency Response Function Models and the Feedback Approach

In cases where the application can't be abstracted into a single, host mass spring and damper, it is useful to introduce the concept of a feedback loop to model the impact of adding a dynamic vibration absorber. It will be shown that the feedback loop concept can also be extended to the much more general case of a finite element model with many degrees of freedom and a whole array of dynamic vibration absorbers.

If we consider just the uncoupled, absorber shown in Fig. 14 and solve for the force, F_r, reacting back on the "floor" boundary at the dashed red line interface as x_r moves in the vertical direction:

$$F_r = c_2(\dot{x}_r - \dot{x}_2) + k_2(x_r - x_2) \tag{14}$$

If we introduce the Laplace variable, s, and solve for x_2, Eq. 14 results

$$x_2 = x_r - \frac{F_r}{(c_2 s + k_2)} \tag{15}$$

Taking the homogeneous equation for the mass, m_2,

$$m_2 \ddot{x}_2 + c_2(\dot{x}_2 - \dot{x}_r) + k_2(x_2 - x_r) = 0 \tag{16}$$

Again, using the Laplace variable, s, substituting Eq. 14 for x_2 and some algebra, we get a frequency response function with units of force per velocity that is governed by the mass, damping and stiffness properties of the absorber:

$$\frac{F_r}{\dot{x}_r} = \frac{\left(c_2 m_2 s^2 + k_2 m_2 s\right)}{\left(m_2 s^2 + c_2 s + k_2\right)} \tag{17}$$

This structural frequency response function is sometimes referred to as mechanical impedance. A useful interpretation of this frequency response function in the context of Fig. 14 is that it is the force that reacts back on the floor when the floor moves up at a unit velocity. With the equation for the force for a unit velocity for the uncoupled device, a VA or a TMD can be added to a single-degree-of-freedom system as a feedback loop as shown in Fig. 15.

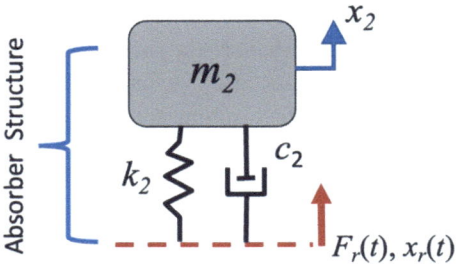

Fig. 14 The absorber system uncoupled from the primary structure and modeling the effect of the primary mass oscillation as base excitation of the absorber mass

3 Analysis and Testing

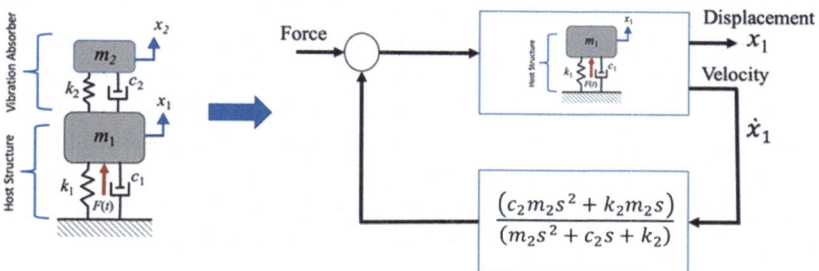

Fig. 15 The schematic of a dynamic vibration absorber is on the left. The block diagram on the right illustrates the equivalent feedback control loop of a dynamic vibration absorber in the Laplace domain

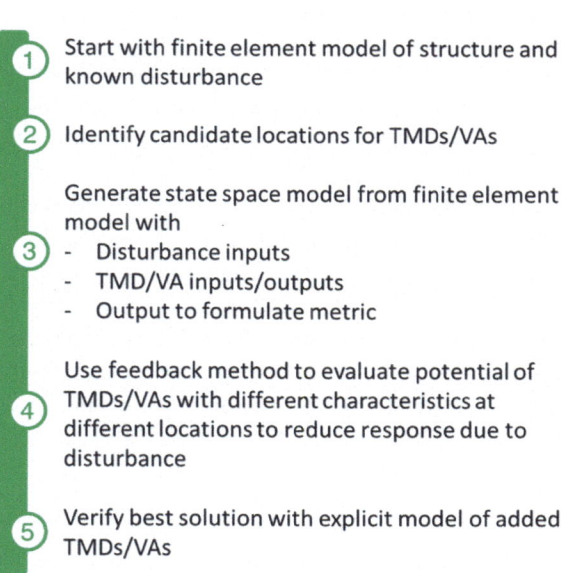

Fig. 16 Steps in analyzing the addition of TMDs/VAs using the feedback method

1. Start with finite element model of structure and known disturbance
2. Identify candidate locations for TMDs/VAs
3. Generate state space model from finite element model with
   - Disturbance inputs
   - TMD/VA inputs/outputs
   - Output to formulate metric
4. Use feedback method to evaluate potential of TMDs/VAs with different characteristics at different locations to reduce response due to disturbance
5. Verify best solution with explicit model of added TMDs/VAs

If we compare the solution to Eqs. 16 and 17 with the closed-loop response using the feedback analogy in Fig. 16, we get exactly the same answer.

3.1.2 Generalization of the Feedback Approach Using Finite Element Modeling

The benefits of the feedback analogy are access to a plethora of feedback control tools to help in coming up with an optimal solution, a simple extension to the more general case of a multi-degree-of-freedom finite element problem and an aid in using test data to design TMDs and VAs.

Making use of the feedback analogy makes modeling the addition of TMDs and VAs to a structure that is modeled with finite elements identical to modeling any other feedback control system that is added to a structure modeled with finite elements. The disadvantage of this approach when compared to explicit models of TMDs and VAs integrated into a structure is, unlike the single-degree-of-freedom feedback analogy, the solutions are not identical. This is because adding an explicit model of a TMD or VA into a structure changes the local behavior of the structure, however slightly. With the feedback analogy applied to a finite element model, part of these local changes may be lost, leading to a less accurate solution. For instance, a modal solution to a dynamics problem is pursued and the mode shapes without the addition of a TMD or VA are used to represent the structure's motion, when we add a TMD or VA with a feedback loop, the local impact on the mode shapes where the device was added is lost unlike with an explicit model. If many modes are included in the solution or residual flexibility is incorporated, this inaccuracy can be minimized. In exchange for a slight loss of accuracy from an explicit model, the feedback approach allows the solution of one, large finite element model to be used to try out many potential locations or multiple combinations of TMDs and VAs. Once an acceptable solution is found, explicit models that include the structure and the added TMDs or VAs can be assembled to verify accuracy of the solution. Figure 16 illustrates the steps involved in employing the feedback analogy to find a solution to a vibration problem on a structure where a finite element model has been constructed.

Exercising Step 3 in Fig. 16 usually involves transitioning analysis tools from a finite element code like NASTRAN to a controls analysis code like MATLAB, using modal superposition. The format of dynamics equations that MATLAB works in is called a state space model, and there are many different state space realizations to recast the dynamics problem in MATLAB. One approach is to first assemble a matrix of mode shapes, Φ, and, using modal superposition, define the response of the structure as

$$x = \Phi \eta \qquad (18)$$

where η are often referred to as modal participation factors, and x is a vector of responses of the structure to an input. Substitution of Eq. 18 into the general expression for the dynamics equation for finite element models

$$M\ddot{x} + Kx = F \qquad (19)$$

along with premultiplication by Φ^T results in

$$\Phi^T M \Phi \ddot{\eta} + \Phi^T K \Phi \eta = \Phi^T F \qquad (20)$$

Recognition of modal orthogonality and mass normalization of modes along with some other mathematical characteristics of finite element solutions results in

$$I\ddot{\eta} + 2\zeta\varpi\dot{\eta} + \varpi^2\eta = \Phi^T F \qquad (21)$$

where modal damping is assumed and $2\zeta\varpi$ and ϖ^2 are diagonal matrices with ζ, a diagonal vector of modal damping coefficients, and ϖ, a diagonal vector of normal mode radial frequencies from NATRAN. The order of Eq. 21 is that of the number of retained modes, much reduced from the order of Eq. 19, which is the number of physical degrees of freedom in the finite element model. If a state variable is defined as

$$z = \begin{Bmatrix} \eta \\ \dot{\eta} \end{Bmatrix} \qquad (22)$$

Equation 21 may be transformed into the state space model format to be used in MATLAB. For instance, if a three-mode model with one force input, a, and one displacement output, s, is assumed, Eqs. 19–22 lead to **A**, **B**, **C** and **D** variables defined by

$$\mathbf{A} = \begin{bmatrix} 0 & 0 & 0 & 1 & 0 & 0 \\ 0 & 0 & 0 & 0 & 1 & 0 \\ 0 & 0 & 0 & 0 & 0 & 1 \\ -\omega_1^2 & 0 & 0 & -2\zeta_1\omega_1 & 0 & 0 \\ 0 & -\omega_2^2 & 0 & 0 & -2\zeta_2\omega_2 & 0 \\ 0 & 0 & -\omega_3^2 & 0 & 0 & -2\zeta_3\omega_3 \end{bmatrix}, \mathbf{B} = \begin{Bmatrix} 0 \\ 0 \\ 0 \\ \Phi_{a1} \\ \Phi_{a2} \\ \Phi_{a3} \end{Bmatrix}$$

$$\mathbf{C} = \begin{Bmatrix} \Phi_{s1} & \Phi_{s2} & \Phi_{s3} & 0 & 0 & 0 \end{Bmatrix} \qquad \mathbf{D} = 0$$

where the subscript a in the **B** vector refers to the mode shape component at the input location and the subscript s in the **C** matrix refers to the mode shape component in the output location. All number subscripts refer to the nth vibration mode. The information for the transformation from a NASTRAN modal analysis to a MATLAB state space model is all available in NASTRAN output files.

Once the state space model has been generated, many MATLAB control analysis tools become available to facilitate the process. For instance, a simple TMD example uses NASTRAN to generate mode shapes and MATLAB tools that analyze feedback control and implement a genetic algorithm to come up with the best implementation of TMDs in step 4.

Step 1

A finite element model of a cantilevered beam is shown in Fig. 17.

The beam is a rectangular tube that is 36" long, 2" wide, 1" tall and 0.25" thick. It is made out of steel and weighs 13 lbs. The disturbance will be applied at the nearest unconstrained node to the base. A modal analysis is performed and the first three modes with motion in the y direction are shown in Fig. 18.

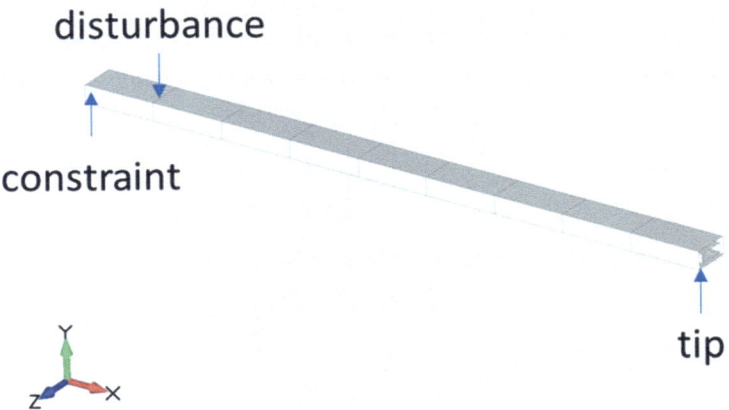

Fig. 17 Illustration of a cantilever beam which forms the structure in this case. The constraint end is usually fixed, and the deflection of the tip is usually the desired displacement to be reduced. The thin lines on the beam are the elements making of the finite element model

Fig. 18 An illustration of the 1st three mode shapes of the cantilever beam and their associated frequencies. This shows locations of the largest amplitudes making good choices for TMD placement

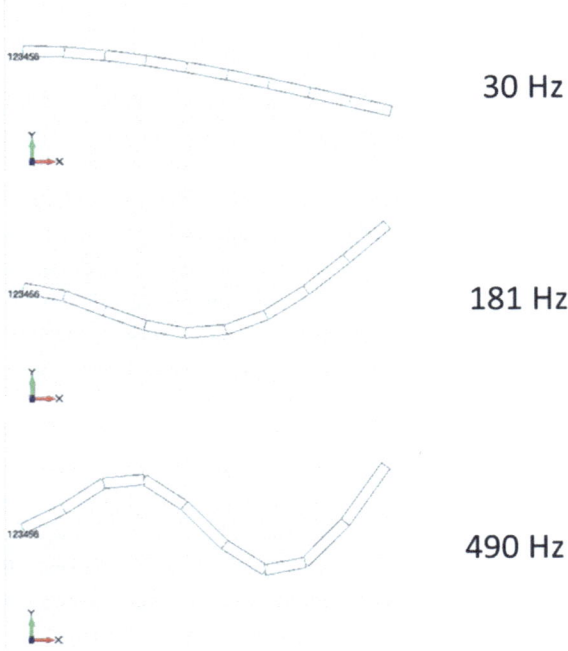

3 Analysis and Testing

Step 2

In this case, all of the unconstrained nodes of the FE model in the y direction will be considered as candidate locations for TMDs. It is expected that for a given vibration mode, locations of highest modal deflection will be the best candidates.

Step 3

A state space model is generated using only the first three modes shown in Fig. 19 in MATLAB using modal superposition [15]. Modal superposition starts with the assumption that a system's dynamic response can be represented by a weighted sum of its natural modes. This premise provides a means of synthesizing frequency response functions in Matlab using mode shapes and frequencies from a normal modes analysis in a finite element code. An example of a frequency response function generated in Matlab that gives the vertical displacement of the tip due to a vertical force applied at the disturbance location near the base is shown in Fig. 19. In this case, the metric is the peak amplitude in this frequency response function, which occurs at the first resonance at around 30 Hz.

Step 4

Single TMD solution using control tool optimization

If our goal is to minimize the peak amplitude at the tip of the beam and we have a constraint that the added TMD moving mass must be ~1 lbs, we can use control design tools to find an optimal solution. Using the feedback approach, we can generate the surface shown in Fig. 20, which gives the maximum amplitude of the first resonance peak in the frequency response function shown in Fig. 21 as a function of uncoupled resonant frequency and damping of the added TMD.

Fig. 19 Frequency response function generated in Matlab using NASTRAN mode shapes showing the first three natural frequencies

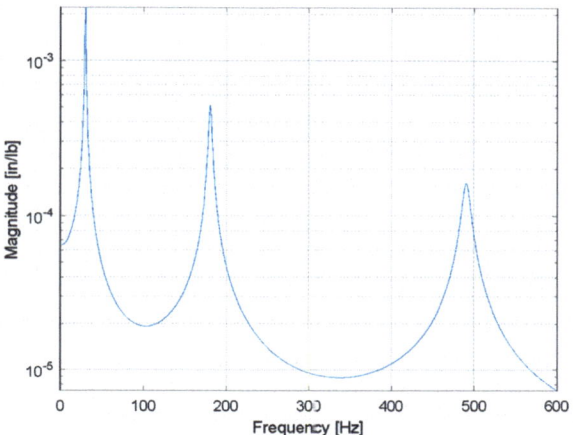

Fig. 20 A surface plot of reduction ratio versus frequency versus damping ratio due to adding TMD generated by using the feedback approach. The red lines show the first peak reduction of the 23 Hz mode with a damping ratio of 0.3

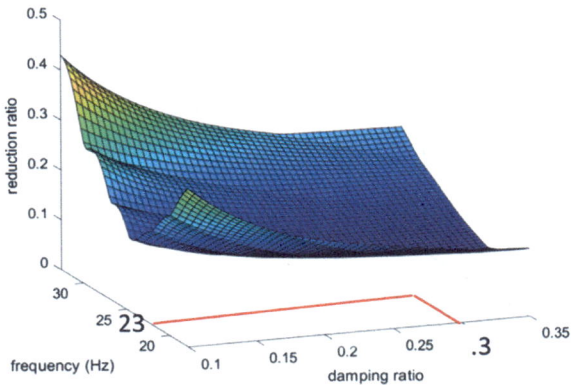

Fig. 21 The frequency response function of the design illustrated in Fig. 20 showing the Impact of adding the TMD to the first mode

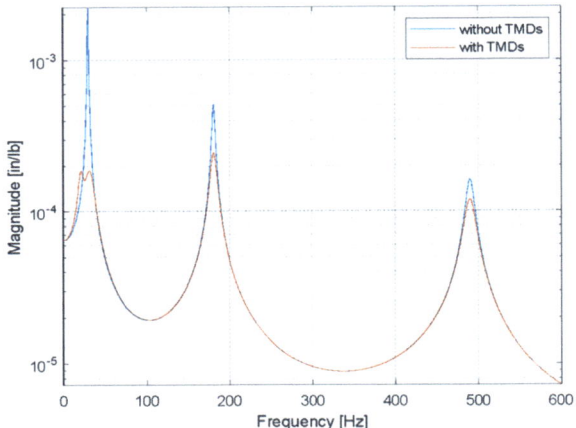

The surface in Fig. 20 has a clearly defined minimum reduction ratio where the TMD has uncoupled resonant frequency of 23 Hz and damping ratio of 0.3. When we look at the impact of adding the optimal TMD as described on the frequency response function, we get the result shown in Fig. 21 with a reduction in the neighborhood of the first peak of almost 90%.

The influence of the correct value for uncoupled damping in a TMD on getting optimal performance may not be obvious. If we take a slice out of the surface in Fig. 18 and look at how the reduction ratio varies with a wider range of damping for an uncoupled TMD frequency of 23 Hz, we get Fig. 22.

The best reduction of over 90% is at a damping ratio of around 30% of critical damping (see reference 1 for definition of critical damping). In close proximity to the optimum, the performance is not very sensitive, meaning an uncoupled damping anywhere in the range of 15–50% of critical would result in close to a 90% reduction ratio. At damping ratios lower

Fig. 22 The variation of peak reduction versus damping ratio by taking a slice out of the surface in Fig. 20 for the uncoupled TMD application. The red line indicates the optimal point

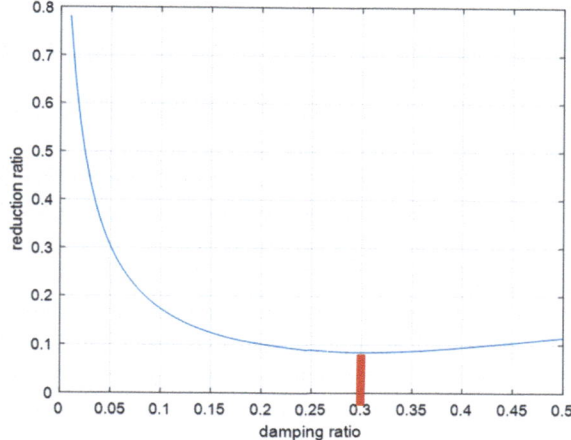

Fig. 23 The variation of peak reduction versus frequency by taking a slice out of the surface in Fig. 20 for the uncoupled TMD application. The red line indicates the optimal point

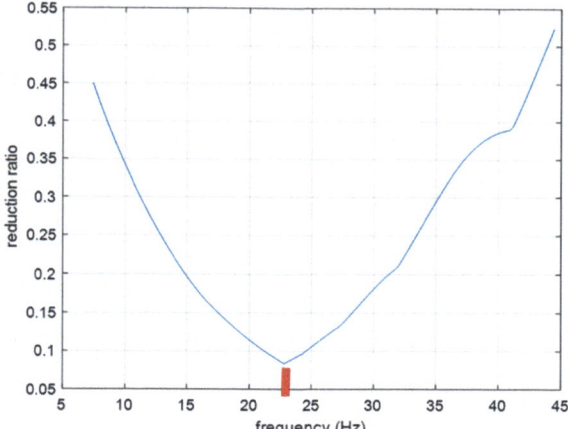

than 15%, performance degrades rapidly with decreasing damping ratio. At damping ratios higher than 50% of critical, performance degrades much more gradually with increasing damping ratio. Since getting uncoupled damping ratio to a high enough number is usually the challenge, the practical problem becomes getting a high enough uncoupled damping ratio and acceptable peak reduction. This rapid degradation of performance at uncoupled damping values significantly below optimum and gradual degradation at uncoupled damping values significantly above optimum is typical of TMD performance.

Looking at the other dimension of the surface in Fig. 20, a plot of peak reduction as a function of uncoupled frequency shows a different trend in Fig. 23.

The best TMD uncoupled frequency value of around 23 Hz is clearly a much more sensitive parameter to get right in order to get good performance, and the sensitivity to getting the value wrong is nearly the same for an uncoupled frequency that is too high as it is for one that is too low. The importance of getting the uncoupled frequency right on a TMD or VA is typical.

For both TMDs and VAs, the sensitivity of getting the uncoupled frequency and damping for the TMD correct is less important for a heavier suspended mass. For aerospace applications, there is usually an aversion to added weight, so making the added mass large to add robustness to uncertainty is usually not an option. A related concept that is not addressed in the analysis presented is the amount of uncertainty in the models of both the structure with a vibration problem and the solution in the form of a TMD or VA. The way to decrease the dependence on a very accurate model and the amount of added weight with both TMDs and VAs is to make a provision in the device to change the frequency to allow in-situ tuning of the device. A provision to tune damping on a TMD is also helpful but not as critical as long as a high damping is achieved. This makes knowing how to test and tune a TMD or VA just as critical as knowing how to design a TMD or VA, especially when added weight allotments are very small.

Another interesting interpretation of the effect of adding the optimal TMD that emerges with the feedback approach is to plot the root locus in Fig. 24 of the system which shows what happens to the complex roots of the structure and the TMD when the loop is closed by adding the TMD to the tip of the beam. The root locus is a common tool used to evaluate the effectiveness of feedback control that is easy to leverage using the feedback approach to TMD design.

The lightly damped mode at the upper right of Fig. 24 that forms the base of the yellow line is the uncoupled structural mode and the heavily damped mode at the lower left of Fig. 24 forms the base of the cyan line. When the TMD is added to the tip of the beam, the two modes are coupled and the result is two damped modes with a damping of 17% for the coupled structural mode and 16% for the coupled TMD mode. Using the root locus tool in

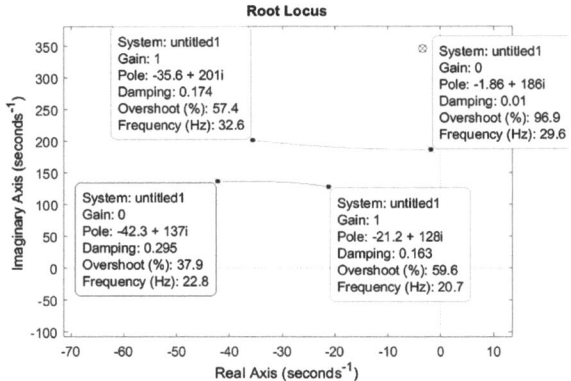

Fig. 24 A plot of the imaginary part of the damped natural frequency versus the real part, call the Root locus of system treating the TMD as a feedback controller

MATLAB could easily lead to an alternate optimization criteria based on the information that is generated by this tool. For instance, one option could be that the coupled modes should have equal damping. Progress toward this goal is easily visualized using the MATLAB rlocus tool.

TMD array solution using optimization
The GA (Genetic Algorithm) is a good choice when a combination of multiple TMDs are being evaluated to satisfy a metric that is dependent on the reduction of a disturbance that is broadband and where the response is dependent on more than one vibration mode. Imposing a penalty on a parameter like added weight makes the GA an especially good choice for optimization, since it can easily handle inputs that are not continuous and TMD parameter choices like weight are discrete and depend on model number. In this case, we will define a metric as a weighted combination of the maximum peak amplitude in the response of the system and the maximum added weight. The TMD mass values at each location are limited to 0.25, 0.5 and 0.75 lb. The solution of no TMD at a specific location is also allowed as a solution in the optimization. Damping values may vary continuously between 8 and 20% of critical and uncoupled frequencies and may vary continuously between 1 and 600 Hz. Figure 25 gives the impact on the frequency response function shown in Fig. 19 for two examples of metrics with different weightings: the one on the right places a much larger penalty on the peak amplitude and the one on the left places a larger penalty on added weight. These results are the result of 40 generations with 100 solutions per generation, making a total of 4000 different solutions. Each solution includes mass, damping and uncoupled frequency at each location making 27 different "genes" for each solution. Each 4000 solution results took about two minutes total solve time using the GA tool. Both results do a good job of reducing the peak amplitude of the largest peak in the frequency response function. For the result on the left, the largest peak gets to a level similar to that of the next largest peak, 2.6×10^{-4} in/lb making an 88% reduction, at an added TMD mass cost of 0.75 lb. The result on the right requires much more added TMD mass, 2.25 lbs, for a modest further reduction of the maximum peak to 1.8×10^{-4} in/lb making a 92% reduction. The added benefit of more reductions in higher modes might make this extra mass option more attractive, but that would require modifying the optimization metric.

Step 5
To form an explicit model to compare to the results using the feedback method, take the example of the optimized 1 lb TMD added to the tip. An explicit model of a TMD with uncoupled frequency and damping of 23 Hz and 30% of critical damping needs to be added to the tip of the beam. A coarse explicit model of a TMD or VA in NASTRAN can be done using cbush elements [16] with the correct stiffness and damping in the axial direction and a very large stiffness in all the other directions. This is illustrated in Fig. 26.

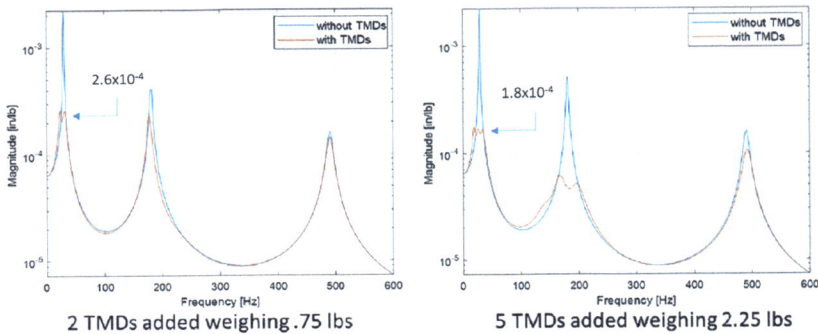

Fig. 25 Magnitude versus frequency plots for two different solutions to minimize different metrics. The left plot corresponds to placing a larger penalty on added weight and the plot on results from placing a much larger penalty on the peak

Fig. 26 The sketch on the left shows the FEM beam model and the call out on the right shows the explicit FEM model of the TMD applied at the tip

In NASTRAN, the mass is specified as having a mass of $\frac{1 \text{lb}}{386 \frac{\text{in}}{\text{s}^2}}$ and no inertia. The cbush element is specified by spring stiffness of $k = 1 \times 10^6 \frac{\text{lb}}{\text{in}}$ in all directions except y (which is the axial direction of the TMD). The stiffness specified in the y direction follows from the uncoupled resonance frequency and the specified mass from the equation that relates stiffness to frequency and mass for a single-degree-of-freedom spring mass damper, $k = (2\pi 23)^2 \times \frac{1}{386} = 53 \frac{\text{lb}}{\text{in}}$. Also, using the equation that relates damping coefficient to damping ratio and mass for a single-degree-of-freedom spring mass damper, we can derive the damping in the y direction as $c = (2\zeta(2\pi 23)) \times \frac{1}{386} = 0.22 \frac{\text{lb}}{\frac{\text{in}}{\text{s}}}$. This completely specifies the single-degree-of-freedom representation of the TMD. A discussion of the equations for c and k can be found in any vibration textbook [2, 17].

The frequency response of the tip of the beam can then be calculated for the beam with and without the TMD for a disturbance near the base as was done before for the feedback model. The result of this calculation is shown in Fig. 27. Comparing Figs. 21 and 27, it is apparent that the solutions are nearly identical. The difference in the problem formulation is that Fig. 21 uses mode shapes of the beam without the TMD to synthesize the frequency response, and the eigenvalue problem becomes coupled when we use the

3 Analysis and Testing

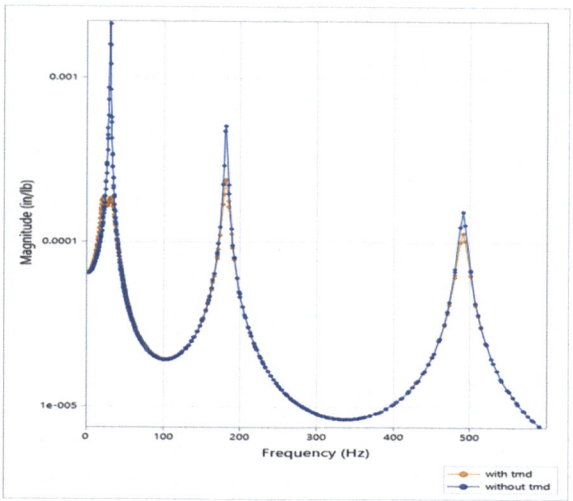

Fig. 27 The frequency response of the tip of the beam calculated for the beam with and without the TMD for a disturbance near the base using the Explicit model of TMD in NASTRAN

feedback command in Matlab. In the NASTRAN version of Fig. 25, the response is synthesized using the mode shapes that include the TMD, and the coupling is apparent in the original eigenvalue problem. The model of the TMD can include much more detail like the actual flexures, moments of inertia for the moving mass and the mass and stiffness of the TMD housing. These will all make the representation of the TMD (or VA) more accurate. However, given the uncertainty of the model the structure will be attached to as well as that of the disturbance path, a coarse model is usually adequate along with provisions in the TMD or VA for in-situ tuning.

The feedback approach to designing an array of TMDs or VAs along with optimization tools in MATLAB open up this technology to problems where these solutions have not been considered. Using the approach outlined, the result gives a point solution that helps define how many dynamic absorbers and how much added weight might be necessary to reduce vibration.

When optimization is carried out on an array of TMDs whether analytically or real-time during installation and tuning, the solution might target specific modes as with the single TMD case illustrated in Fig. 19. If added weight is not an important factor in minimizing the metric, a broadband effect might be achieved by incorporating a TMD with a relatively large suspended mass and large damping at the low end of the frequency range of interest. Physically, this can be interpreted as isolating the TMD mass and reacting back against the structure with the relatively lossy suspension of the TMD. In most of the TMD results shown, this effect is also observed by noting that lightly damped modes at frequencies higher than the mode that is primarily damped, also exhibit increased damping when the TMD is incorporated. The concept of reacting against an isolated inertial mass is exactly how an inertial actuator works [18]. In this extreme case, a broadband

TMD treatment could be realized using one or several inertial actuators tuned to a relatively low frequency with an electrical resistance across the leads that results in the correct mechanical damping.

3.1.3 TMD Array Solution Using Collocated Frequency Response Functions

When correlated finite element models for complex structures and/or good definitions of the disturbance path are not available, the optimized solution is unlikely to perform as well as predicted. In the case of a complex structure with lightly damped modes and an ill-defined disturbance path, an array of TMDs can be designed using only collocated frequency response functions at the points where vibration reduction is required and at the points where TMDs can be added. If possible, using the point where vibration reduction is required as a point where a TMD can be added is desired but not required. It is also necessary to define the frequency range over which vibration reduction is required as well as any frequency-dependent weighting that might help to choose vibration modes to damp. A successful TMD array solution derived only from collocated frequency response functions assumes.

(1) The vibration modes observed at a selected frequency at the point where vibration is to be reduced are the same as those observed at the points where TMDs are to be added.
(2) Adding damping to lightly damped modes observed in the weighted frequency range specified at the point where vibration is to be reduced will reduce the collocated response at that location.
(3) Some or all of the modes observed in the collocated response at the point where vibration is to be reduced will also be important in the response at the same point from a noncollocated disturbance.

Taking the correlated model of a portion of a 767 aircraft used to explore adaptive noise cancellation in large aircraft [19], mass and inertia elements can be added to represent four seated passengers as is illustrated in Fig. 28.

In this case, a seated passenger is abstracted as a mass and inertia that is attached to the aircraft floor using a rigid element at four locations for each mass/inertia representation.

If it is assumed that an axial TMD that acts in the vertical (y) direction may be mounted at each of the four floor locations for each of the four "passengers", there are 16 possible TMD locations available. If the goal is to reduce broadband vibration at the four passenger locations without any additional information about the disturbance path, the approach outlined in Fig. 29 may be followed. This approach is analogous to sequential loop closure [20] in feedback control systems. Each time a tuned mass damper is added to the system, the system changes and a new measurement is required to find the lightly damped vibration mode that results in the largest amplitude at the point where vibration is to be

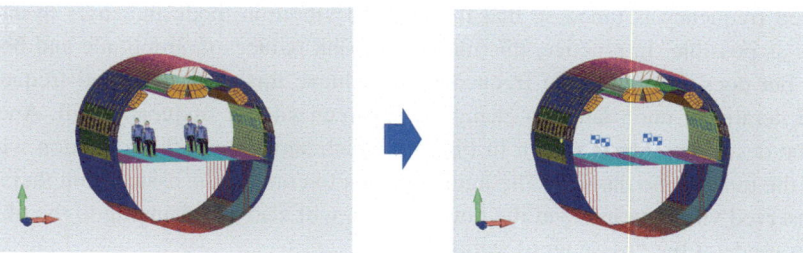

Fig. 28 Finite element model of portion of 767 aircraft barrel section with passengers and seats represented to illustrate points where vibration is to be reduced. Note the yellow dot on the right-hand side figure which represents the disturbance location

reduced—in this case the passenger location. The overall goal is to add damping to as many lightly damped modes as possible at the passenger location in the frequency band of interest.

For the case of passengers subjected to vibration, ISO 2631-1 [21] gives a frequency dependent weighting to assess caution zones for seated passengers subjected to acceleration in the vertical direction. The frequency response function description can be applied as the weighting in the process described in Fig. 29 to add damping to vibration modes that are most likely to amplify unhealthy vibration.

The process in Fig. 29 was followed using the model illustrated in Fig. 28. The same approach can be used on hardware using measured, collocated frequency response functions. Figure 30 shows the progression of the weighted, collocated frequency response function at the first "passenger location" as the four tuned mass dampers are added to each floor location using the feedback method. The tuned mass dampers are assumed to have a suspended mass of 4 lbs and an uncoupled damping ratio of 8% of critical. The

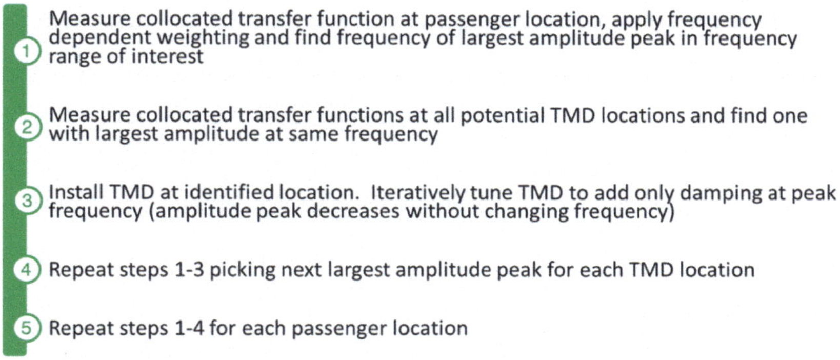

① Measure collocated transfer function at passenger location, apply frequency dependent weighting and find frequency of largest amplitude peak in frequency range of interest

② Measure collocated transfer functions at all potential TMD locations and find one with largest amplitude at same frequency

③ Install TMD at identified location. Iteratively tune TMD to add only damping at peak frequency (amplitude peak decreases without changing frequency)

④ Repeat steps 1-3 picking next largest amplitude peak for each TMD location

⑤ Repeat steps 1-4 for each passenger location

Fig. 29 Steps to tuning an array of tuned mass dampers when only collocated frequency response functions are available

uncoupled frequency is tuned so that the targeted vibration mode increases in damping as much as possible. In practice, this means the peak reduces in amplitude and becomes broader but does not change in frequency. To achieve this, the uncoupled frequency is adjusted iteratively until damping change is apparent without frequency shift. An example of the desired tuning is shown in Fig. 30, which shows each floor location with and without the tuned mass damper. The figure appears to confirm the assumption that adding to an observed vibration mode at the floor location will also add damping to the observed vibration mode at the "passenger location".

After all 16 locations were tuned in the manner described, a disturbance location was selected on the left-hand-side forward, mid skin. This location is marked in Fig. 28 with a large, yellow dot on the right-hand side illustration of the finite element model. This location is approximately coincident with the location where a pressure input was assumed in an earlier study using the same model [19]. A normal force was applied at the disturbance location and the resulting, normalized weighted response at each of the four, passenger locations is shown in Fig. 31 with and without TMDs.

The reductions in the weighted, RMS acceleration responses due to the disturbance are 43, 41, 35 and 35% respectively. These reductions are of similar magnitude to the reduction in response from the highest to lowest acceleration response for four hours of

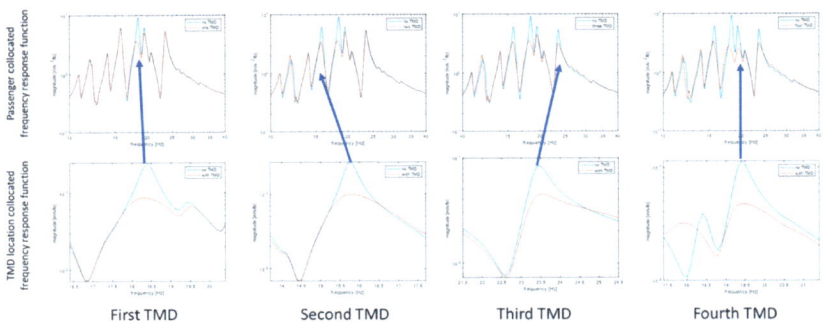

Fig. 30 Progressive impact on collocated frequency response functions at TMD locations (lower figures) and passenger location (upper figures) as TMDs are added at all four locations

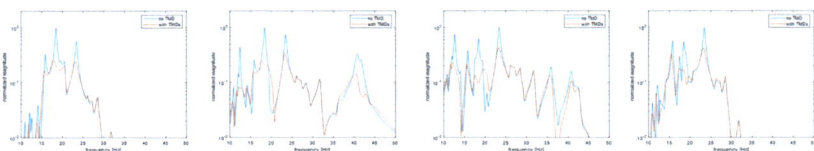

Fig. 31 Normalized, weighted response due to disturbance from noncollocated disturbance location marked by yellow dot in Fig. 28

exposure in ISO 2631-1, with the magnitude of the caution zone spanning from approximately 60–100% of the maximum allowable acceleration. This means if a passenger were being exposed to vibration near the maximum of the range of the caution zone, adding tuned mass dampers in the manner described could reduce loading to a level that is below the lowest level of acceleration in the caution zone.

The authors have used the same approach on large, complex aerospace structures with multiple, lightly damped vibration modes in a critical frequency range and without a well-defined disturbance path. The approach was first used with the finite element model to define the number and size of the TMDs required and later used on the actual aerospace hardware with an observed RMS acceleration reduction that was consistent with the earlier analysis and close to an order of magnitude less with TMDs over the desired frequency range.

3.1.4 The Dynamic Vibration Absorber as a Metastructure

Using more than one vibration absorber or tuned mass damper can be advantageous when more than one resonant frequency is bothersome or when one would like a range of frequencies to be neutralized. Most of the original work was academic and examined the analysis of plate vibrations with two or more vibrations absorbers attached. Likely the first effort to model this was Snowden [19], later formalized and generalized by Nicholson and Bergman [20] in a series of papers. The basic concept was to attack each mode of resonance with a single DVA. The area of mechanical metamaterials is another application of multiple DVA type devices. In this case the goal is to create ranges of frequencies where no resonance occurs, similar to creating band gaps in acoustics metamaterials [21]. The idea is based on repeated lattice-like structures as indicated in Fig. 36, where each "cell" is an absorber. Most mechanical metamaterials are composed of a ridged frame with a distribution of repeated structures internal to it, each of which acts as a small vibration absorber. The major difference between traditional absorber considerations and the mechanical metamaterial approach is that the absorber designs considered in the previous sections are added on after manufacturing, whereas the metamaterial approach is a complete redesign of a structural component. Subsequently, metastructures are neither TMDs or VAs (Fig. 32).

The goal of mechanical metamaterials (also called metastructures) is to produce a frequency response with a range where no resonance can occur by shifting frequencies such that there is a substantial gap between two frequencies corresponding to the frequency range of any applied forces, thus avoiding resonance. Because such devices are a complete redesign, the possibility exists of providing a new structure having the same mass as the original structure but with a substantial frequency gap, avoiding resonance. The following simple example shows this by examining the longitudinal vibrations of a hollow rectangular cross section as pictured on the left of Fig. 33.

Constant Mass Example A square cross section channel indicated on the left in Fig. 33 is redesigned as a metastructure as indicated on the left by redistributing the total mass of

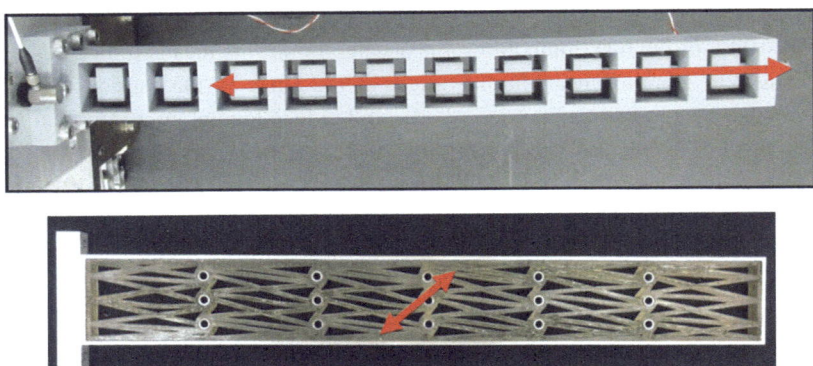

Fig. 32 Two examples of mechanical metamaterials. The top version is constructed for suppression of longitudinal vibrations and the bottom photo is for transverse vibration where the black dots serve as absorber mass and the longerons as the stiffness elements

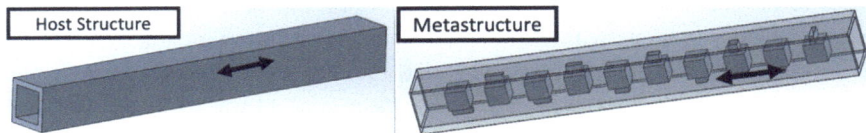

Fig. 33 The host structure to be protected is given on the left. The redesigned system as a mechanical metastructure by redistributing the mass is shown on the right

the structure. Both structures have the same mass, and the metric of interest is reducing the tip deflection in the longitudinal direction.

The first design is to keep each spring and mass of each absorber the same and redistribute the mass to minimize the total energy in the tip deflection. The second design uses inverse eigenvalue theory to place all the system natural frequencies above 600 Hz, simulating a scenario where driving frequencies are all below 500 Hz. One of the design parameters is the number of spring-mass systems to use. In the case shown, only two absorbers are used, and the new structure is modeled as a five degree-of-freedom system. The response of the three structures, all of the same mass, is given in Fig. 34.

The frequency response in Fig. 34 shows that the metastructure design splits the first natural frequency much like an attached TMD would, but for no added mass. However, using the inverse procedure which assigns a different stiffness to the two absorbers, the frequency response shows no resonance below 600 Hz. The impulse response of the three systems also shows improvement for the two metastructure designs. Such an approach may be useful in some circumstances. As mentioned earlier, this redesign for vibration

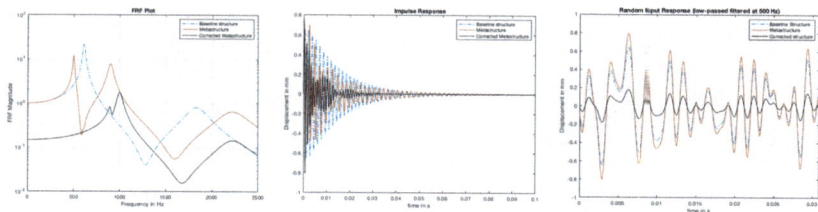

Fig. 34 The left shows the frequency response of the three systems: base structure, metastructure and the corrected metastructure placing the lowest frequency above 900 Hz for a 10-resonator metastructure system. The middle plot shows the time response of each system to an impulse excitation [22]. The plot on the right shows the response to random input low-pass filtered at 500 Hz. The input is at the fixed end of the structure and the responses shown are of the tip (free end)

suppression would ultimately have to consider all other design factors such as strength, toughness and fatigue.

Three-Dimensional Example

An additional potentially useful aspect of metastructures is the ability to absorb in three directions. Figure 35 illustrates a 3D printed metastructure column in which each cell contains absorbers that can move in three directions, as illustrated in Fig. 32.

The experimental results shown in Fig. 35 were made at the Army Research Laboratory on their multi-axis shaker. The structure was not optimally designed but put together to investigate what was possible in multiple directions. The spring stiffness and masses illustrated in Fig. 36 can be designed to produce a variety of frequency responses.

In summary, the metamaterial approach has the ability to create structures of the same mass but designed for vibration suppression and to create structures that absorb vibration frequencies in three directions (torsional, longitudinal and transverse). The technical approach is relatively new and has not yet revealed a specific application.

3.2 Testing Methods

A treatment for a TMD or VA can be designed based on experimental measurements only. The process is similar to a standard modal test but the goal is not to measure mode shapes and damping, but to fit a math model to modal data that can be used to predict and design a vibration treatment. This goal makes the process one of system identification as opposed to modal test, but the tools used, shakers, hammers and accelerometers, are those that belong to the modal test category. Another benefit of the feedback approach introduced in Sect. 4.3 is that it easily extends to experimental design of both TMDs and vibration absorbers.

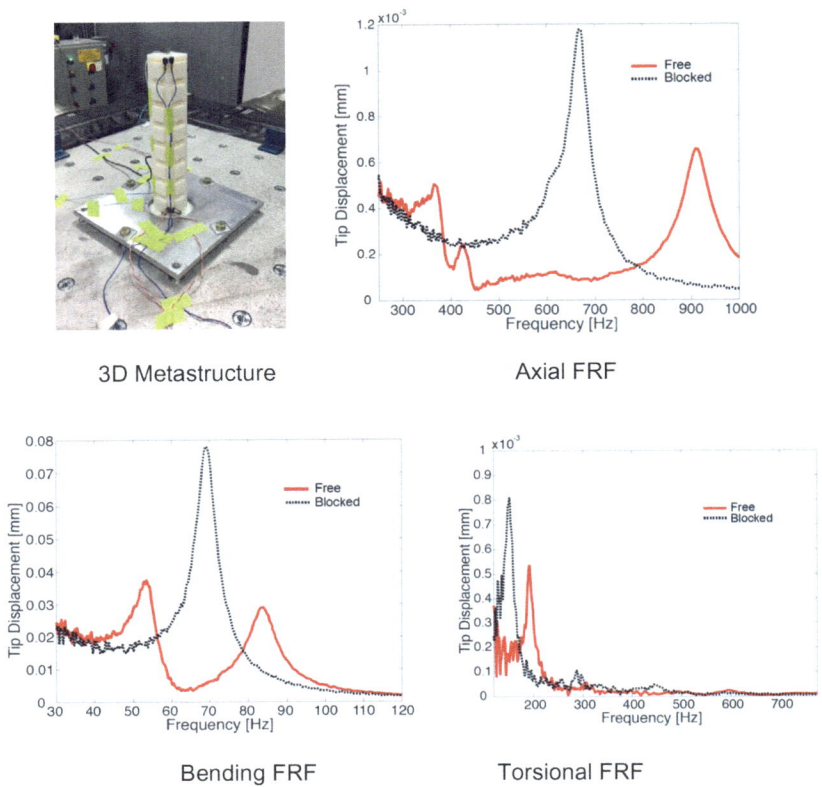

Fig. 35 A 3-D printed metastructure is in the upper lefthand corner followed by the measured frequency response functions of the tip for each of the 3 directions do to a base excitation from a 6-axis shaker

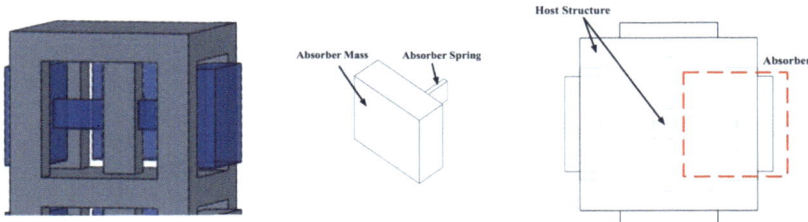

Fig. 36 The unit cell of the metastructure in Fig. 35 configured to absorb vibration in multiple directions: torsional, longitudinal and transverse

3.2.1 TMD Design Using Modal Test

Single location, single mode

In a large, aerospace structure, where the goal is to reduce the response of a single, lightly damped mode and access is limited to a single location, we start by measuring a collocated frequency response function. In this case, we use a modal hammer (3 lb) and a collocated accelerometer. The units of the measurement are in acceleration per force, but because the expression for the term that will be fed back to represent the tuned mass damper has a velocity input, the measurement should be converted to units of velocity per force by integration (dividing by frequency in radians per second). For a single mode fit this makes it convenient to convert the measurement to units of velocity per force via integration (admittance). In this case, the single mode fit was carried out iteratively by varying the frequency and damping of a second order model and minimizing the RMS error in the neighborhood of the peak. For more complex models, the MATLAB system identification toolbox can be used. Figure 37 shows the measured admittance frequency response function along with a single mode fit.

The mass term in the single mode fit is a useful quantity as it sets bounds on the required suspended mass for the tuned mass damper. In this case, the mass is 2006 lb mass. Similar to the analysis, the typical ratio range of suspended mass to collocated starts at 1/100 and goes to 1/10. In most cases, suspended masses at a single location in excess of 10 times the fit mass give diminishing returns. Figure 38 shows the variation in amplitude reduction for this ratio range assuming the uncoupled frequency of the added TMD is 95% of the frequency of the mode to be suppressed and the uncoupled damping of the added TMD is 20% of critical.

Fig. 37 Measured admittance transfer function from large, aerospace structure along with single mode fit to first mode as curve fit by iteratively by varying the frequency and damping of a second order model and minimizing the RMS error around the first peak

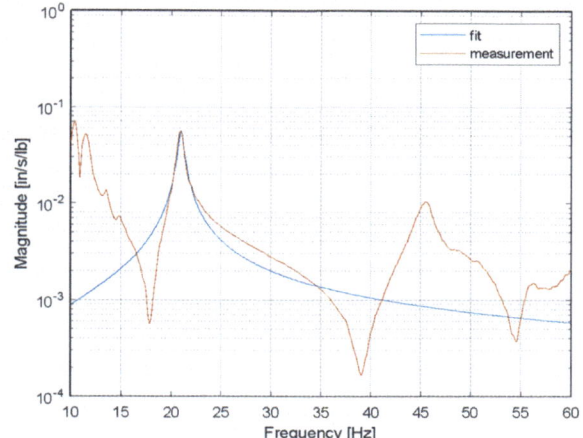

Fig. 38 The variation in amplitude reduction of the peak showing the dependence of reduction of on TMD mass for the uncoupled mode for $\zeta = 0.2$

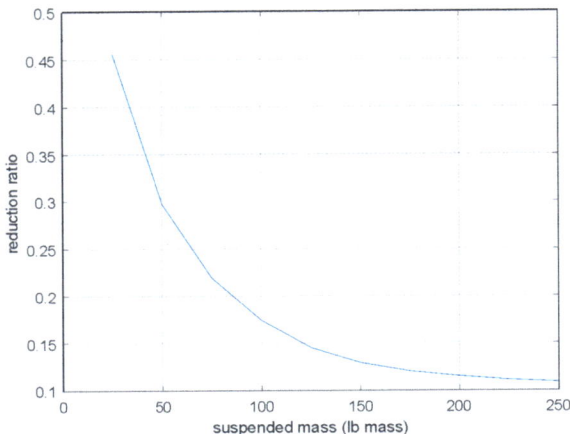

Observing Fig. 38, it appears that there are diminishing returns on reducing the response at a suspended mass of around 150 lb mass. At 200 lb mass, the aforementioned factor of 10 is reached and further reduction is negligible.

There are three variables to adjust for our single location device and all of the variables are at least weakly coupled.

– Uncoupled mass of the TMD
– Uncoupled damping of the TMD
– Uncoupled frequency of the TMD

If none of variables are constrained by what is available for purchase or manufacture, the easiest way to find an optimal solution is an optimization that minimizes a metric that represents reduction. A metric that calculates the amplitude over a band that extends to either side of the peak is often more robust than one that just calculates reduction at the original resonant frequency, especially if the frequency content of the disturbance is unknown. Other metrics might include the root locus method that tracks the damping of the coupled modes or the area under the curve in the admittance frequency response function where the curve extends across a band that includes the peak. If the disturbance is known, the predicted response using the frequency response function makes the best metric.

For the example in question, assume a limit on the suspended mass of 150 lb mass. The variation in the metric of peak reduction with uncoupled damping and uncoupled frequency to determine the best values to minimize the metric can then be calculated as in Fig. 39.

3 Analysis and Testing

Fig. 39 A surface plot of reduction ratio versus frequency versus damping ratio for the uncoupled case

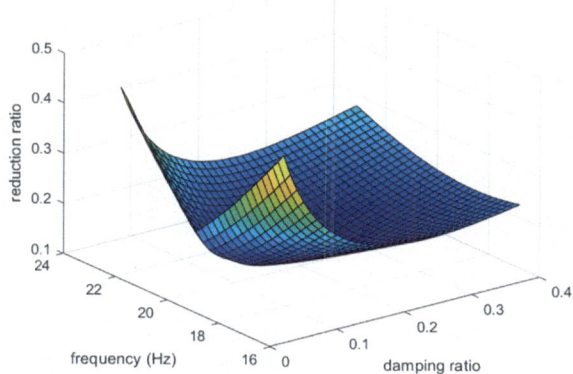

Fig. 40 A plot of magnitude versus frequency showing the reduction of peak of large, aerospace structure at single location using TMD with 150 lb-mass suspended mass

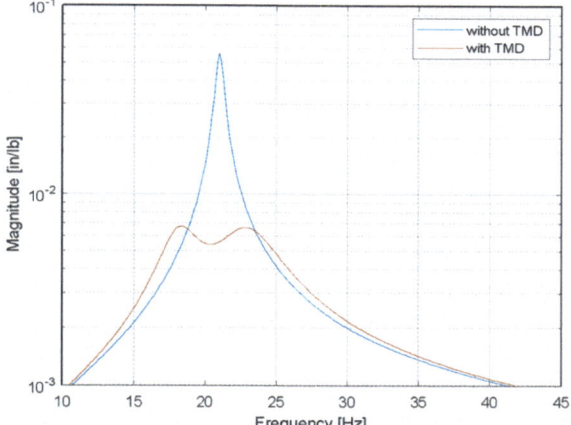

The resulting predicted performance at the surface minimum is shown in Fig. 40.

If the installed device doesn't match expected behavior, mechanisms that allow for in-situ tuning are very helpful. For the single-degree-of-freedom example, tuning rules are shown in Fig. 41. In either case, the tuning objective is to move the relatively damped peak closer in frequency to the lightly damped peak, increase coupling and decrease response. In the case on the left of Fig. 41, the heavily damped peak is at around 14 Hz and stiffening the TMD or removing mass will cause it to increase in frequency. In the case on the right in Fig. 41, the heavily damped peak is at around 33 Hz and softening the TMD or increasing mass will cause it to decrease in frequency.

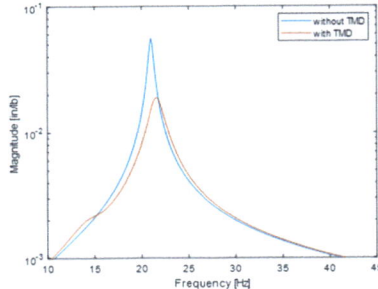
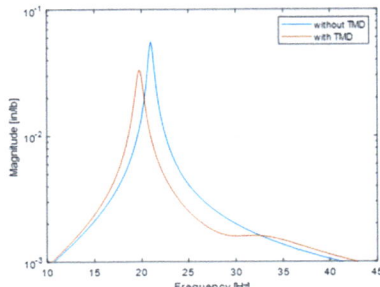

Try stiffening TMD or decrease mass   Try softening TMD or increase mass

Fig. 41 In-situ tuning rules by looking at various cases. The plot on the left moves the maximum amplitude frequency to the right whereas the plot on the right moves the maximum amplitude frequency to the left

The previous single-degree-of-freedom, single location example covers many of the situations where TMDs are employed. To summarize, Fig. 42 illustrates the steps involved using the feedback method.

Multiple locations, single mode
In this example, the large aerospace structure had two locations available to suppress motion with the constraint that each may have a suspended mass of 72 lb mass and that each device must have the same tuning to simplify manufacturing. The model representing this case gets a little more complicated. We need to not only identify the frequency and damping of the mode we want to suppress but also the mode shape components at the two locations and a disturbance location if it is different from the TMD locations. This can be accomplished using standard modal test software and modal testing techniques. In this case we want to identify a model with.

- Disturbance location (input and output)
- Rear TMD location (output)
- Forward TMD location (output)

A modal hammer at the disturbance location with accels at the disturbance, rear and forward locations was used to collect the frequency response functions in units of admittance (in/s/lb) in Fig. 43.

This data was used to synthesize a three input, three output modal model using only the mode of interest at around 22 Hz. The resulting model was then used to optimize the uncoupled damping and frequencies of identical devices incorporated at each location with a maximum suspended mass of 72 lbs. The resulting with and without TMD measured

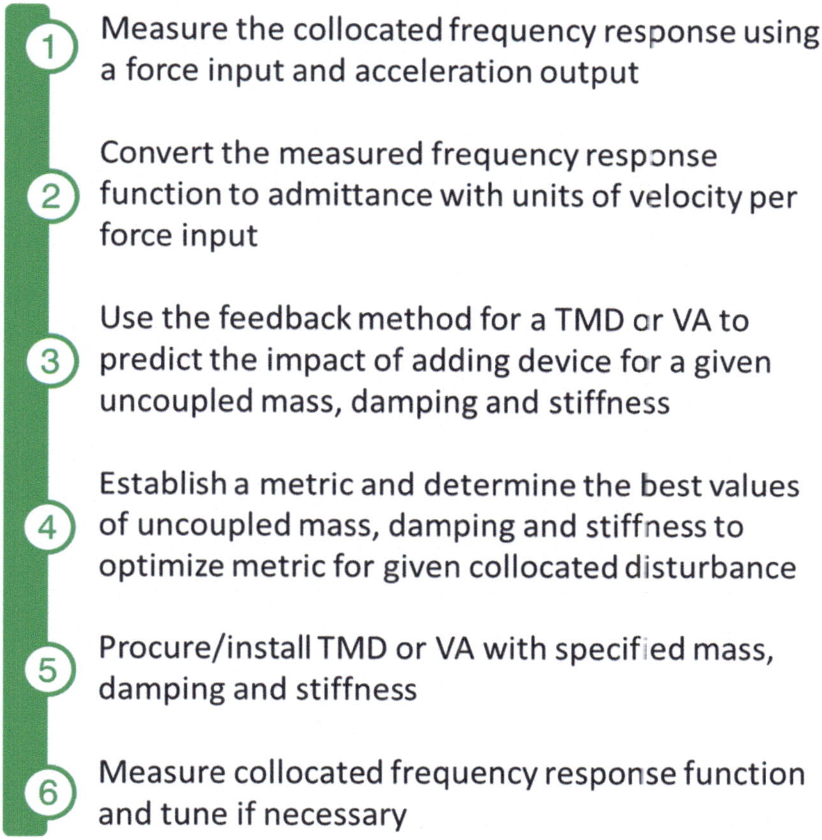

3 Analysis and Testing

1. Measure the collocated frequency response using a force input and acceleration output

2. Convert the measured frequency response function to admittance with units of velocity per force input

3. Use the feedback method for a TMD or VA to predict the impact of adding device for a given uncoupled mass, damping and stiffness

4. Establish a metric and determine the best values of uncoupled mass, damping and stiffness to optimize metric for given collocated disturbance

5. Procure/install TMD or VA with specified mass, damping and stiffness

6. Measure collocated frequency response function and tune if necessary

Fig. 42 A step by step set of rules to design a single mode, single location TMD/VA

result is shown in Fig. 44. The reduction shown was sufficient to solve the vibration problem on the aerospace structure. The mode is effectively eliminated as a means of amplifying disturbance in that frequency range.

The result in Fig. 44 involved tuning because the measured data with no TMDs installed and the resulting synthesized modal model were used to derive the best uncoupled damping and frequency given a mass constraint. When the disturbance is not collocated, multiple sensor locations are important if further tuning is required when the TMDs are installed. The previous approach outlined for a single–input–single–output system becomes untenable in this situation. If in-situ tuning is attempted, it needs to be done while tracking a multiple sensor metric and a noncollocated input because a local reduction in response does not necessarily guarantee a global reduction.

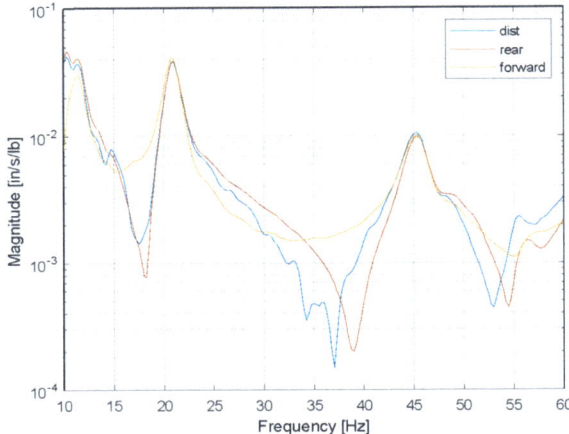

Fig. 43 A measured plot of response magnitude versus frequency (admittance) for the presented example problem

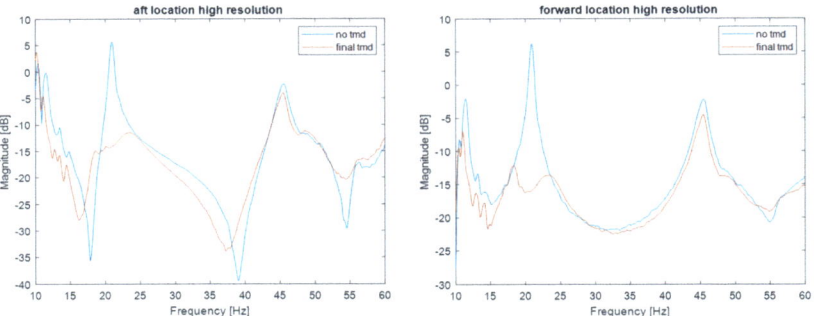

Fig. 44 Measured plots of response magnitude versus frequency for adding a TMD at two different locations to illustrating the impact on a large aerospace structure

Extending the approach of using the experimental technique outlined from multiple inputs and outputs and suppression of a single mode to suppression of multiple modes is just a matter of changing the metric and the order of the model that is used to determine the TMD parameters. For instance, a TMD array could be designed to significantly dampen both the mode at around 30 Hz the mode at around 45 Hz, but the synthesized model would have to be modified to include the mode at 45 Hz and a metric that considered the area under the curve from approximately 25–50 Hz would need to be minimized.

3 Analysis and Testing

3.2.2 VA Design Using Modal Test

When there is a periodic disturbance and the goal is to limit its response at some known location, a VA is the right solution. As an example, take the case of a disturbance applied to the bed of a pickup truck shown in Fig. 45, where the goal is to suppress the response at a known, noncollocated location, in this case, the top of the wheel well.

The mathematical fit to the one input, two output system is done using the MATLAB system identification toolbox. Unlike the TMD example, with VAs, measuring and fitting models to multi-input, multi-output measurements where the disturbance does not excite a resonance is required. The tools used for this kind of mathematical fit can be similar to those used for the TMD example, but to accurately model off-resonance behavior, it is usually required to use more general system identification tools as opposed to single degree of freedom models. This is because when we install a VA at a given location, there is no guarantee that the overall vibration will reduce anywhere except at the VA location. There is also no guarantee that the residual response away from resonance is dominated by one mode. Usually, the location of interest is at some distance from the VA. This

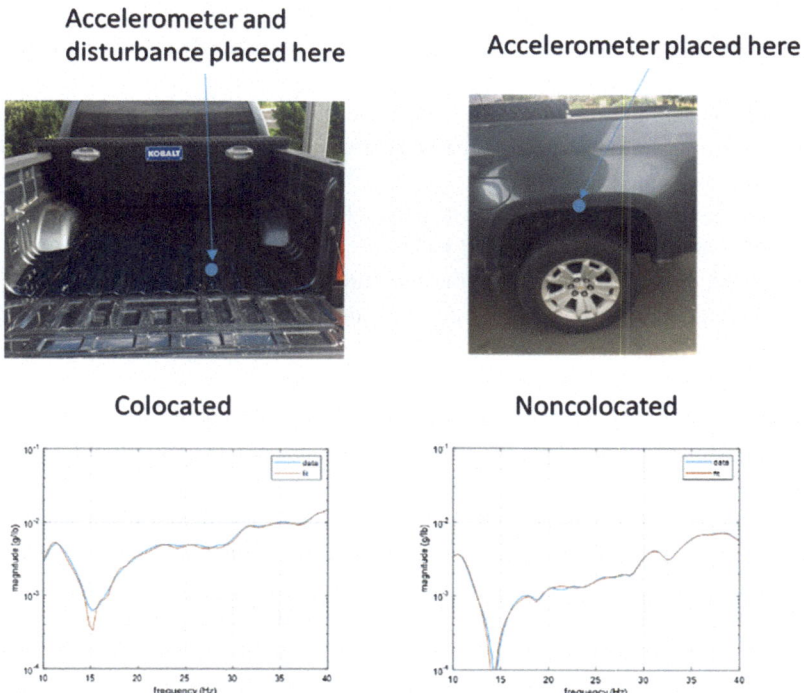

Fig. 45 The top left photo shows a vibration disturbance in a truck bed, the top right shows the location to be suppressed, the bottom left shows the frequency response at the bed and the bottom right shows the frequency response at the wheel well

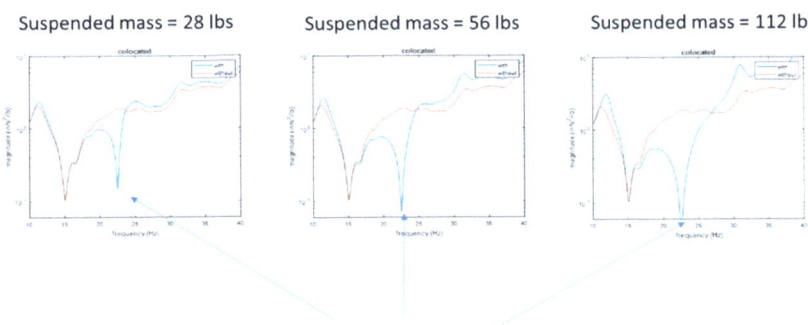

Fig. 46 The frequency response plots of three different VA designs with increasing mass (left to right) showing how the reduction increases with added suspended mass

requires at least an input for a noncollocated disturbance, inputs for one or multiple VAs, outputs for one or multiple VAs and outputs at the locations where the response is to be reduced.

Once a math model is available, the feedback tools introduced in Sect. 4.3 can be used to design a vibration absorber for a given frequency disturbance. Taking the example of a disturbance frequency of around 22 Hz, we can close our feedback loop for three different suspended masses, resulting in a progressively greater attenuation at 22 Hz, as shown in Fig. 46 for the collocated location.

Figure 47 shows a VA with the 56 lb mass glued to the bed of the truck and along with an inertial shaker at the collocated location.

Once the VA is incorporated into the structure, we can repeat our modal test and compare to predictions at both the collocated and noncollocated locations for the VA with a 56 lb suspended mass as shown in Fig. 48.

For the case of the VA, this approach does a good job of predicting the impact of adding a vibration absorber to the relatively complicated structure of the truck. We can also "play" a disturbance at 22 Hz using the inertial shaker with and without the vibration absorber as shown in Fig. 49.

Although there is a dramatic reduction of the tone generated by the shaker at the noncollocated location with the addition of the VA, the reduction is not as large as the predicted or measured results shown in Fig. 49 for the noncollocated location. Both the frequency domain results in Fig. 49 show more than a $20\times$ reduction at 22 Hz, while the time domain result using the shaker in Fig. 47 is more like a $4\times$ reduction. The most likely reason for this discrepancy is that the shaker can't be placed exactly at the same location as the VA and the disturbance path from the VA to the noncollocated position is not the same as that from the shaker to the noncollocated position. The difference in the path is due to the relatively compliant truck bed. This difference would be less dramatic for a more rigid connection between the shaker and the VA and/or a more similar path

Fig. 47 The inertial shaker and TMD glued in bed of truck

from each to the noncollocated position. Alternately, adding an additional input to the collected data at the disturbance location for the math model that was used may also have resulted in a prediction that was closer to the measured reduction and may have even suggested a slightly different tuning of the VA to minimize the transient response. Finally, this difference illustrates that it is important when collecting the "without VA" data to incorporate any feature of the VA to be used that might locally stiffen the installation area. This was done in this example by collecting the initial data set with a plate with equivalent stiffness to the VA housing installed at the intended VA installation location.

Colocated

Noncolocated

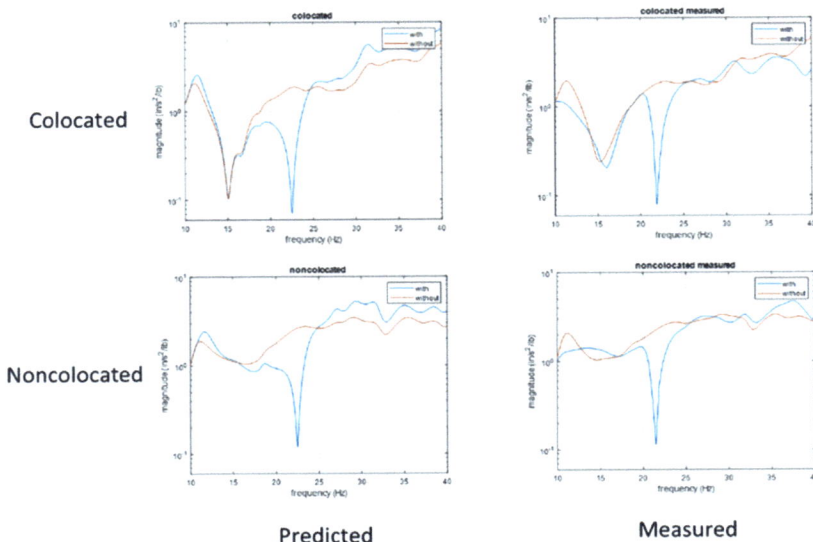

Predicted   Measured

Fig. 48 Predicted and measured frequency response functions at two locations, the wheel well (noncollocated) and the truck bed (collocated)

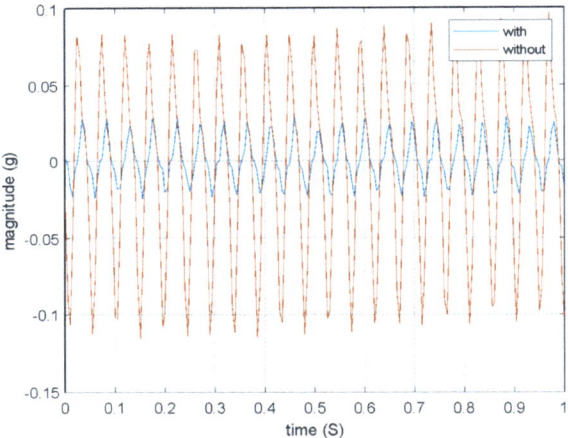

Fig. 49 Transient behavior with and without TMD at the noncollocated (wheel well) location

4 History and Lessons Learned

Dynamic vibration absorbers are used in a variety of applications throughout the world. Below are some examples of TMDs and VAs being used to limit vibration in aircraft and non-aircraft environments.

4.1 Vertical Tails

TMDs have been investigated for use on the vertical tails of some fighter jets to counteract vibration during high angle of attack maneuvers, which cause unsteady flow thus exciting vibration modes of the structure as shown in Fig. 50.

An example of a representative TMD that could be used in this application is shown in Fig. 51 [8]. The frequency response of a design that incorporates such dampers is significantly lower than that of a design without the dampers, as shown in Fig. 52.

Fig. 50 An F15 encounters unsteady flow causing the twin tails to vibrate. See also Fig. 2

Fig. 51 Representative TMD for vertical tail problem (Image courtesy of Moog)

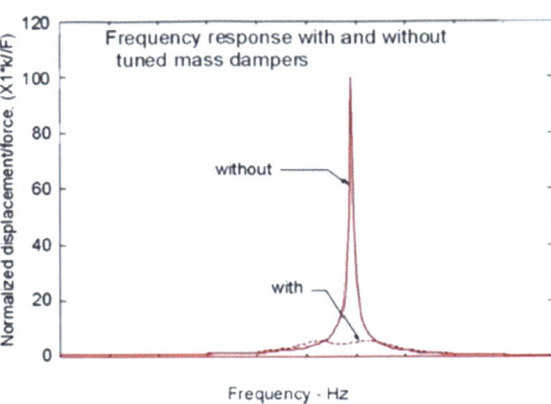

Fig. 52 Tuned mass damper frequency response (Image courtesy of Moog) illustrating the effect of adding TMD on the magnitude

4.2 Self-tuning Mass Dampers

The need to provide a means of tuning TMDs after installation is illustrated in Figs. 40 and 41. A VA might also require slight tuning but this only involves adjusting frequency and can be accommodated with a relatively simple addition or subtraction of moving mass. Since TMD performance relies on getting both damping and frequency correct, its use benefits from a provision for self-tuning, where changes in damping and frequency

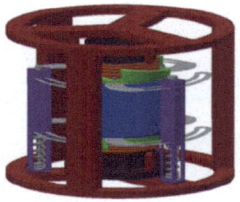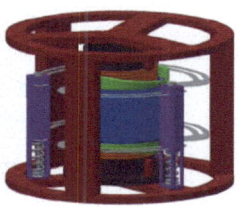

1. Initial configuration with flexures at mid length. Solenoids locked.
2. Mass moved up and rotated clockwise. Solenoids released.
3. New configuration with flexures at max length. Solenoids locked.

Fig. 53 The mechanical approach to self-tuning TMD using solenoids (purple) and flexures

can be made remotely and even autonomously if necessary. The use of a voice coil as the loss mechanism in a TMD enables at least two potential means of self tuning:

- Mechanical adjustment of flexure stiffness and damping adjustment using added variable resistor [23]
- Active feedback loop to collocated differential sensor for active adjustment of stiffness and damping as well as centering loop for ultra-low frequency operation [24]

Self tuning becomes critical when an array of TMDs require in-situ tuning. On the aforementioned, large aerospace structure with 32 TMDs, ~75% of the effort in a quite extensive test of flight hardware was removing and reinstalling TMDs in order to make adjustments.

The mechanical adjustment approach to a self tuning TMD recognizes that the voice coil loss mechanism can also be used as an actuator as illustrated in Fig. 53.

The purple components in Fig. 53 are solenoids that lock or release circumferential, variable length flexures. The operation illustrated starts with step 1 and the solenoids locked at the middle of the flexures. In step 2, the solenoids are released at the same time as a voltage is applied to the voice coil which pushes the suspended mass up so that the green ramps make contact with red bearing surfaces at the top of the housing. This causes the mass to rotate clockwise and extends the flexure to its maximum length. The final step 3 is to lock the solenoids. The frequency of the TMD is decreased by this operation. The operation can also be repeated with the opposite voltage to decrease the length of the flexures. The voice coil can also be used as a sensor and actuator to determine the frequency of the suspended mass by applying a dynamic excitation and monitoring the current draw of the voice coil. Once the correct frequency is determined, the damping can be changed by varying the added resistance on the coil. This approach is described in reference 19 but is not available in a COTS device. This approach could also be applied to a VA but it is unlikely that a voice coil would be added to a VA just for the purposes of self-tuning.

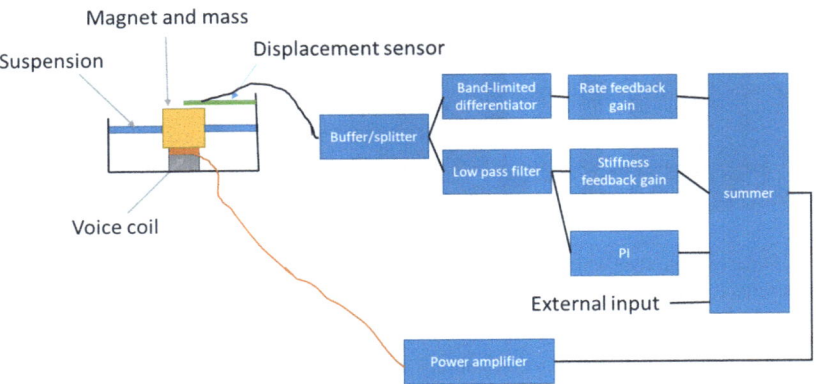

Fig. 54 A schematic of an active feedback system to form a self-tuning TMD

The active feedback approach to a self-tuning TMD also requires the addition of a differential sensor between the housing and the suspended mass. It is illustrated in Fig. 54.

In this approach, the differential rate and displacement are fed back to increase or reduce the stiffness or damping terms. This approach has been demonstrated in hardware to show a range of adjustment of about ±50% of the nominal, passive frequency and a similar range of adjustment for damping. The motivation for this device was to change the uncoupled frequency and damping values of a TMD that was deeply embedded in hardware and needed to change values depending on operational conditions. This approach also offers the added benefit of closing an active positioning loop at a very low bandwidth on the suspended mass to keep it in the middle of its range. This allows very low frequency operation even with gravity acting in the axial direction of the TMD. This approach has been shown to achieve resonant values of ~0.1 Hz for a suspended mass. The active version of the self-tuning TMD is described further in the referenced patent [24] but is not available as a COTS device.

4.3   Taipei 101

TMDs are often used in tall buildings like the Taipei 101 in Taiwan [25]. Designed to withstand typhoon winds and earthquake tremors, the Taipei 101 was designed with a 660 metric tons (728 short tons) steel pendulum as its Tuned Mass Damper suspended from the 92nd floor to the 87th floor. The pendulum sways to offset the building moving during strong gusts of wind (Fig. 55).

Unlike most buildings that incorporate TMDs to reduce the response to wind and earthquakes, Taipei 101 invites people to see and understand the technology by creating a viewing area to observe the primary TMD.

4 History and Lessons Learned

Fig. 55 The Taipei 101 building (center) and its large TMD. The sketch on the left shows the location of the TMD and the photo on the right is looking down on the pendulum mass (see Fig. 11 for the placement of the dampers)

4.4 MD-80/DC-9 Aft Cabin

The MD-80, DC-9 and Boeing 727 all share tail mounted engines. In all three cases, passengers seated near the engine installation as shown in top view in Fig. 56 from Lang [6].

These aircraft have all incorporated vibration absorbers to reduce periodic vibration at certain flight conditions. Lang [6] documents one such effort for the MD-80. Vibration absorbers were placed around the barrel of the fuselage as is shown in Fig. 57 as well as on the engine mounts.

The approach is effective as is shown in the flight data in Fig. 58.

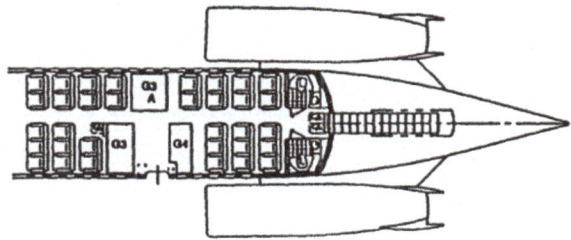

Fig. 56 The top cut away view of the MD-80 showing the engines proximity to passengers

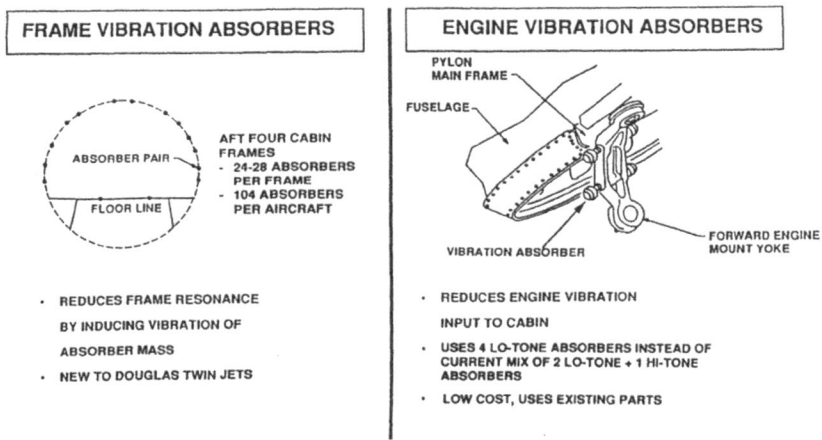

Fig. 57 Two different approaches to vibration absorbers to reduce engine noise in the cabin. See also Fig. 1

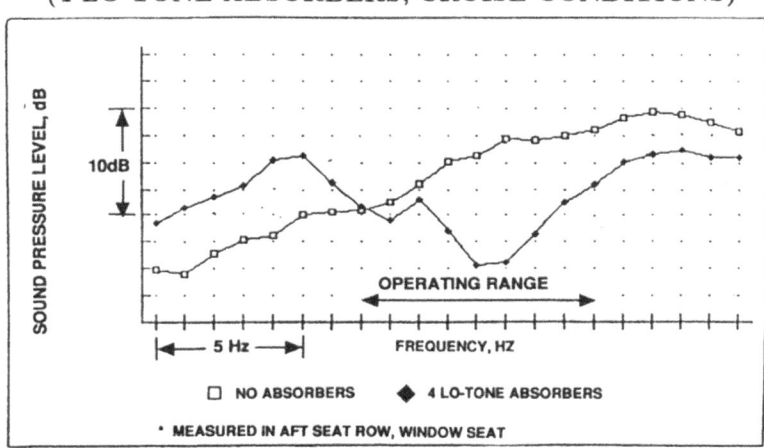

Fig. 58 A closeup of Fig. 1 showing the effect of adding VAs to the MD 80 by measuring sound pressure levels

It is important to note that the treatment has been designed to match the operating range and that tuning is very important. If the uncoupled frequency of the absorbers were higher or the operating frequency range had extended lower, the added VAs would have made the sound problem worse instead of better.

4.5 Passive VAs on Helicopters

Although modern CH-47's use active devices to suppress periodic vibration caused by the rotors, Bramwell [26] gives some details the previous self-tuning VA that was used along with a similar approach on the Sea King helicopter that makes use of its battery as a suspended mass. Figure 59 illustrates the large battery in green suspended on a spring suspension in blue. The reference also details passive vibration absorbers on the rotor head.

Some challenges in implementing a vibration absorber on a helicopter like the Sea King are that the rotor frequency is not necessarily constant, necessitating the need for an adjustment. In the case of the semi-active absorbers, this led to an actuation system which changes the absorber resonance as a function of rotor speed. This requirement for a constant adjustment to maintain performance naturally leads to the use of active vibration cancellation, which is currently implemented on the CH-47 and many other multi-rotor platforms. Active vibration cancellation has the advantage of requiring less volume and added mass for the same effective cancellation force as a passive treatment but has the disadvantage that it requires power and additional control electronics for operation.

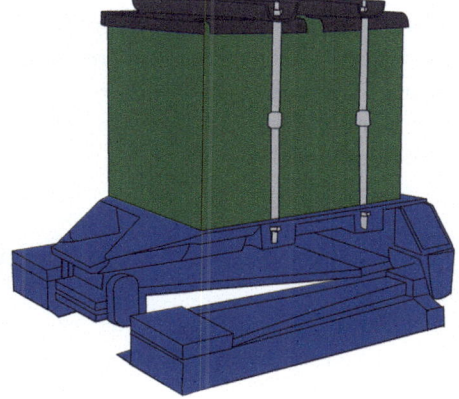

Fig. 59 An illustration of the vibration absorber used in the Sea King helicopter

5 Hands on Hardware Description and Instructions

Simple demonstrations of a tuned mass damper and a vibration absorber can be assembled using common materials available from a hardware store and a 3d-printed flexure. Slightly more sophisticated demonstrations can be accomplished with the addition of easy-to-acquire piezoceramic pickups and an oscilloscope.

5.1 Host Structure

Both the tuned mass damper and the vibration absorber demonstrations require a cantilevered beam, host structure. For the demonstration described the following materials were used,

- 0.125″ x 0.75″ × 17″ aluminum beam
- Clamp[1]
- Dental mirror
- Laser pointer[2]

The clamp that forms the cantilevered boundary condition is important to achieve a lightly damped host structure that will clearly change behavior when a damper or a vibration absorber is attached. A clamp that will attach to a stout table and provide a tapped hole to attach the beam is preferred but not required. Another important element of a simple demonstration is a laser pointer directed off a mirror attached to the end of the beam. Figure 60 shows the hardware the host structure as described. The holder for the laser pointer is 3D printed out of a polymer. The print fill can be found at holder.stp [24]. The laser/mirror combination is a great way to amplify the motion of the tip of the beam. Figure 61 also demonstrates how a small motion of the beam translates into a larger motion of the laser spot projected onto the wall. A greater distance to the wall results in a greater amplification.

5.2 Addition of a Tuned Mass Damper

Once the host structure has been constructed, and the lightly damped, amplified vibration of the beam is clearly observable upon tapping the beam, the next step is to design a tuned mass damper to reduce both the amplitude and duration of its response. The beam shown in Fig. 61 had a first mode of 14 Hz and measured damping of ~1% of critical. Instructions on how to measure these quantities will be included at the end of this chapter,

---

[1] SLOW DOLPHIN Photography Super Clamp w/1/4" and 3/8" Thread Clip.
[2] Petsport USA Laser Chase II Pet Toy.

Fig. 60 The materials used to build a VA for demonstrating the effect. From left to right: a clamp, a laser pointer and a dental mirror. The beam to be suppressed is above the ruler which indicates the scale

Fig. 61 An illustration of how the laser and dental mirror amplify motion of beam so the effect can be easily observed

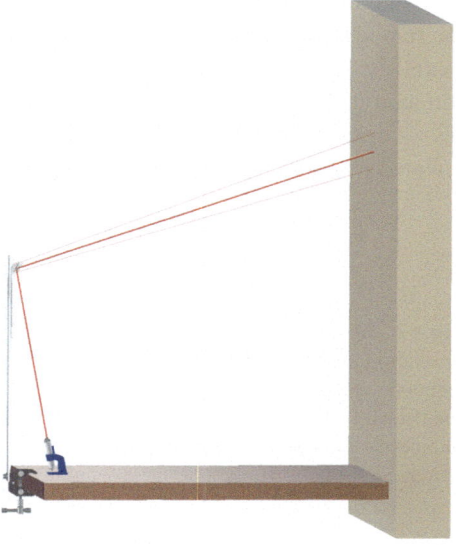

but it is not necessary to get the numbers exactly correct to design the damping in the TMD. For the first cantilever mode of the beam, an estimate of equivalent mass of an idealized, single-degree-of-freedom system for use in designing a TMD can be assumed to be equal to the modal mass at the point of maximum deflection [27]. In this case, the beam weighs 0.15 lbm, so the equivalent single-degree-of-freedom mass is 0.037 lbm or a fourth of the total mass. This forms the basis for an estimate of the required mass of the tuned mass damper. A similar answer can be derived by measuring or synthesizing the collocated frequency response function of the displacement for a given force at the tip of the beam and fitting an equivalent spring mass. This approach gives a single-degree-of-freedom mass of 0.04 lbm, is more general as it can be applied to a measured, collocated frequency response function anywhere on the structure. Either method gives an estimate of the range of masses necessary to add as the moving mass in the tuned mass damper. In the case of the beam, one should expect the moving mass of the tuned mass damper to be between 1/50th and 1/10th of the equivalent, single-degree-of-freedom mass. If the point selected is not the point of maximum of the modal vector, the modal mass estimated using either approach will be larger by the square of the ratio of the maximum modal vector amplitude divided by the modal vector amplitude at the selected location [27]. For a given suspended mass in a TMD, this means the ratio of suspended mass to calculated modal mass is significantly smaller away from the point of maximum deflection. Since the performance of the TMD tends to increase with an increase in this ratio, it is very important to select a location where modal deflection is at or near a maximum to minimize the added weight of the TMD.

Using the methods described in Sect. 3.1.1, Fig. 62 gives a prediction of the effect on the frequency response function at the tip with a tuned mass damper at the heavy and light extremes. The magnitude is unity normalized to easily assess the reduction associated with different TMD treatments. In this case, the damping of the tuned mass damper is set to 5% of critical for both cases and the uncoupled frequency is adjusted until the resulting coupled peaks are approximately equal in amplitude.

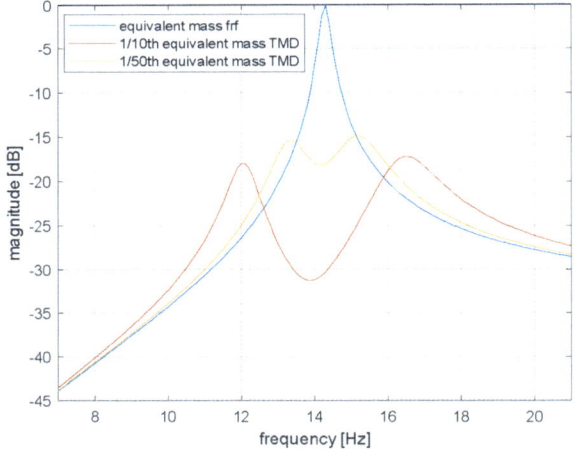

Fig. 62 An analytical prediction of the effect on the frequency response function at the tip with a tuned mass damper at the heavy and light extremes

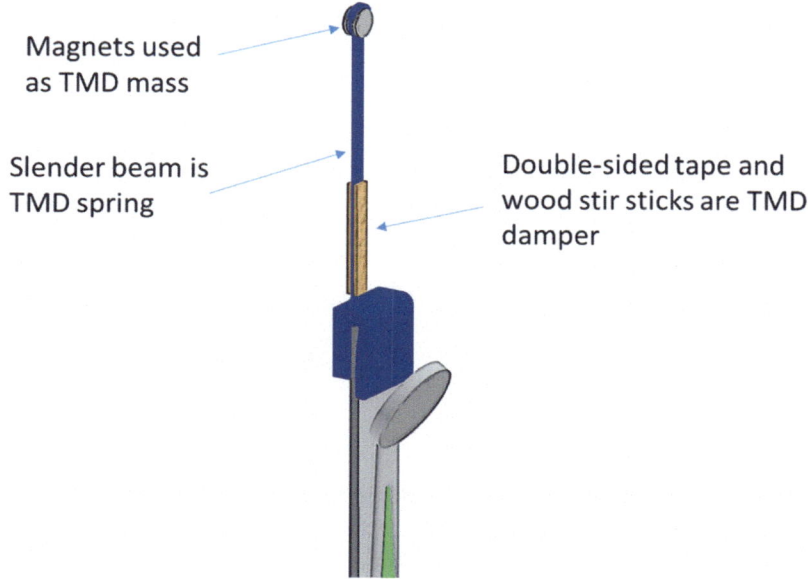

Fig. 63 The tuned mass damper configuration for the beam experiment in Fig. 61

The tuned mass damper to be added to the host structure is shown in Fig. 63. The tuned mass damper consists of.

- plastic (blue) part is printed out of plastic and is available as a solid model [28]
- added mass (0.004 lbm) of the 0.3125″ D × 0.04″ magnets (1/10th of the equivalent mass)
- double sided tape (0.002″ thick 3 m VHB [29]). This material makes a good damping material, although that is not its original intended consumer purpose.
- wood stir sticks cut to size (~1.25″ × 0.24″ × 0.06″)

The tuned mass damper was applied to the host structure and the length of the wood sticks as well as the number of layers of double-sided stick tape were iterated until the behavior of the beam was clearly damped when compared to the host structure. As a final adjustment, the magnet masses can be moved on the beam to fine tune the frequency of the damper. For this example, a small (0.0004 lb) accelerometer was located an inch below the tip of the beam with a small, modal hammer on the other side of the beam as a disturbance. Similar but less quantitative testing can be accomplished with the laser as the sensor and a fingertip as the disturbance input. The original and with TMD behaviors of the beam were measured and are shown in the frequency and time domains in Figs. 64 and 65. Figure 64 can be compared to the prediction in Fig. 62. The amplitude reduction

Fig. 64 The measured frequency response with and without TMD of the demonstration experiment

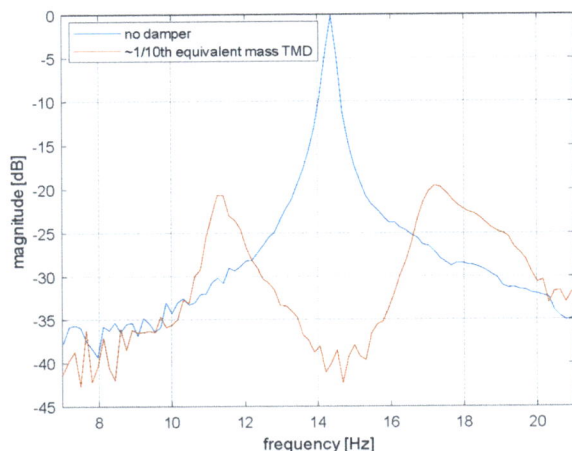

and general predicted trend of adding the TMD in Fig. 64 are very similar for the case of the 1/10th equivalent mass TMD in Fig. 62.

Some potential reasons for not achieving better agreement with prediction in order of expected impact are.

- Added device mass larger than modeled
- Tuned mass damper not modeled explicitly
- Miniature accelerometer has high noise floor
- Linear approximation of relatively large motion
- Input force assumed in frequency response function estimate
- Damping of secondary system not measured
- Measurement not at exactly same place as predicted

Given these potential sources of uncertainty and the observation that the prediction predicted less than the measured reduction, the difference in amplitude reduction of around 2.5 dB seems very reasonable and agreement between predicted and measured behavior is very good. Note the accuracy of the design depends on the sophistication of the equipment and modeling.

To summarize, Fig. 66 gives a step-by-step procedure for designing and testing the tuned mass damper described in this section. In addition, a video is provided [28] to illustrate the steps in Fig. 66.

5 Hands on Hardware Description and Instructions

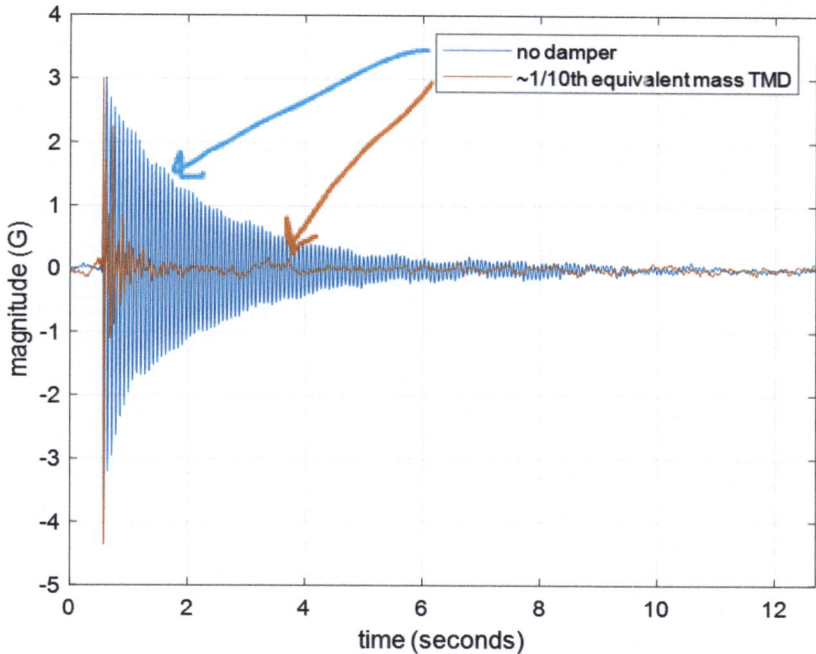

Fig. 65 The measured transient response with and without TMD of the demonstration experiment

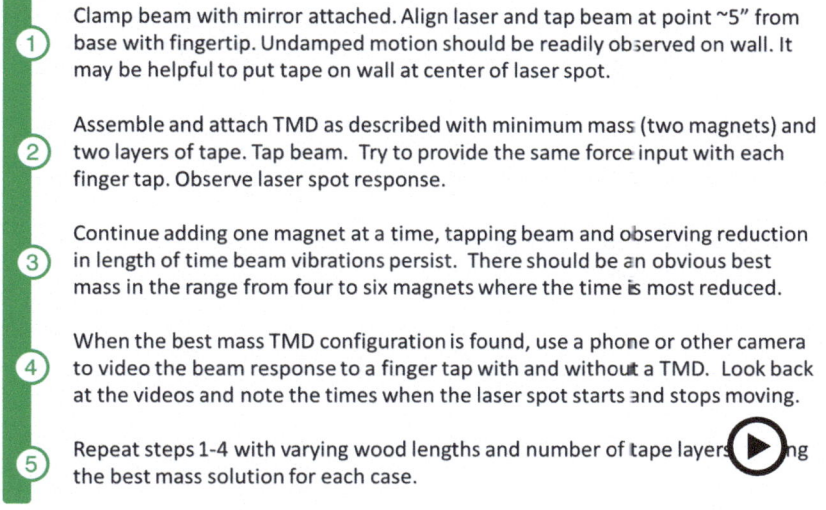

Fig. 66 The steps to completing TMD demonstration experiment (▶ https://doi.org/10.1007/000-a6j)

5.3 Addition of a Vibration Absorber

The added mass penalty associated with the addition of a vibration absorber does not follow the simple rule of a tuned mass damper. Unlike the goal of adding damping to a specific vibration mode of a host structure, which can be represented by an equivalent mass with some specific fraction of the host structure mass, the vibration absorber is designed to reduce the dynamic response of the structure away from resonance. The response away from resonance is made of a combination of many vibration modes. The required added mass must be determined by modeling or measuring the host structural response due to and input force at the location where the vibration absorber will be attached. Figure 67 shows the predicted steady-state, normalized response for a given input force at a frequency below the resonance of our host structure along with the predicted reduction in response due to adding a tuned mass damper with an uncoupled resonance of around 12.4 Hz and with various suspended masses. The largest reduction is 34 dB, for the 0.0064 lbm VA.

The tuned mass damper from Fig. 67 can be used without the damping element (stir sticks and double-sided stick tape) as shown in Fig. 68. In this case, the magnets mass can be moved up or down to match the uncoupled frequency of the vibration absorber to the frequency of the disturbance.

Figure 68 also shows the disturbance generator used in the vibration absorber demonstration. It is a DC motor with two eccentric masses that are used to make a very inexpensive haptic actuator. The holder for the motor shown is provided as a solid model [28]. When the motor spins, a centrifugal force is generated that follows the rotation of the motor. The voltage applied to the motor determines the spin rate of the motor, which determines the frequency of the excitation. To select a vibration frequency, a variable

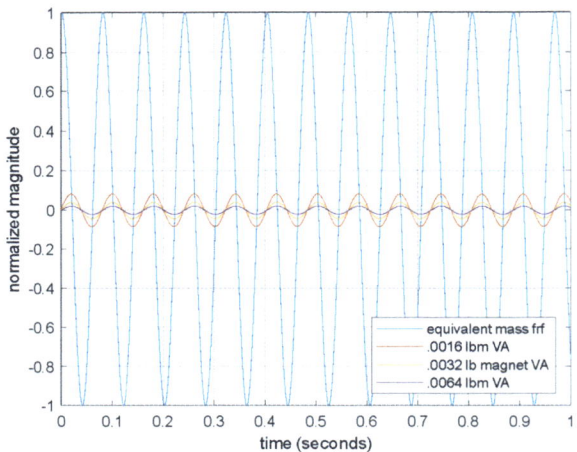

Fig. 67 The predicted steady-state, normalized response for a given input force at a frequency below the beam resonance along with the predicted reduction in response due to adding a VA

5 Hands on Hardware Description and Instructions 63

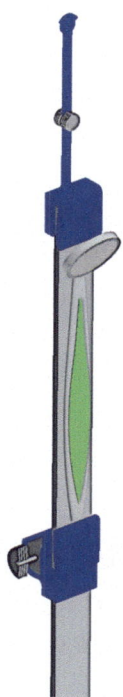

Fig. 68 An illustration of the beam with a disturbance generator (gray piece near the bottom) and a vibration absorber

power supply can be used but can also be accomplished with a battery and a combination of resistors in series with the motor. In the case of the assembly shown in Fig. 68, the motor produced a 12.4 Hz disturbance to the beam. Figure 69 shows the impact on the beam's steady-state response with and without the vibration absorber. In this case, the reduction of 28 dB with the added 0.0064 lbm magnet shows reasonable agreement with the predicted 34 dB reduction for the same added mass. The reasons that the match is not more accurate are similar to the reasons for disagreement between prediction and measurement for the tuned mass damper.

To summarize, Fig. 70 gives a step-by-step procedure for designing and testing the vibration absorber described in this section. In addition, a video is provided (28) to illustrate the steps in Fig. 70.

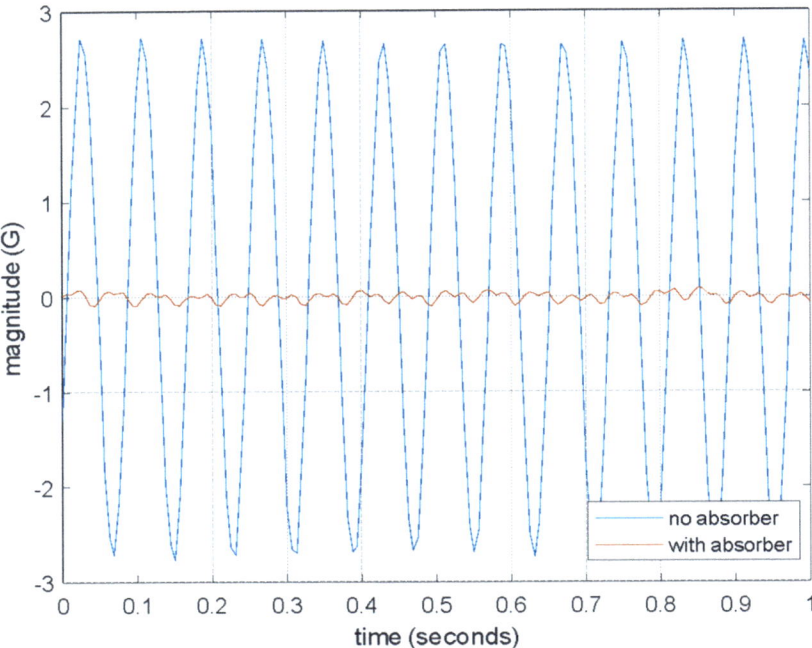

Fig. 69 The measured time response of the beam with and without absorber showing the reduction in amplitude

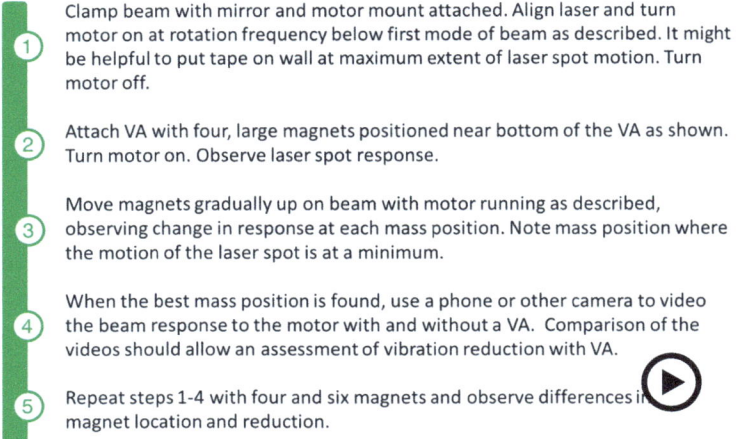

Fig. 70 A step-by-step procedure for designing and testing the vibration absorber described in this section (▶ https://doi.org/10.1007/000-b6d)

5.4 Inexpensive Ways to Measure Vibration

Oscilloscopes
There are at least three different types of inexpensive oscilloscopes that are easy to use:

- Network connected using smart phone
- USB connected using computer
- Handheld miniature

All have advantages and disadvantages. While both the network connected and the USB connected devices facilitate data collection and storage for use in other software like Matlab, the handheld miniature does not require a separate device to see the resulting measurement. Most options also have a provision for converting time domain collected data to the frequency domain to observe the impact of adding a dynamic vibration absorber. Figure 71 shows examples of all three oscilloscopes.

Sensors
The two types of sensors that are easy to incorporate and allow more quantitative measurement of vibration are MEMS accelerometers and piezoceramic guitar pickups. Both are available for around $10. Figure 72 shows both sensors incorporated on the host structure. Since observation of lightly damped vibrations in the "without damper" case are desired, it is preferable to use high gauge (30 AWG) wire to connect sensors and to offload the wire wherever possible to avoid impeding beam vibration. Figure 72 shows an example of a beam with a mems accelerometer[3] at the tip and a small piezoceramic guitar pickup[4] at the base. Piezoceramics can be used as both sensors and actuators on beams. An example of an active control system that is relatively simple to build is given in reference 25.

Both sensors when combined with oscilloscopes, readily identify the first mode of the beam. Figure 73 shows normalized frequency and time domain results of the response of the beam due to an impulse input.

The section provided a simple way to design and build a demonstration of tuned mass damper system for a simple cantilevered beam using relatively inexpensive tools. However, the idea is easily extended to complex structures as indicated in some of the examples cited.

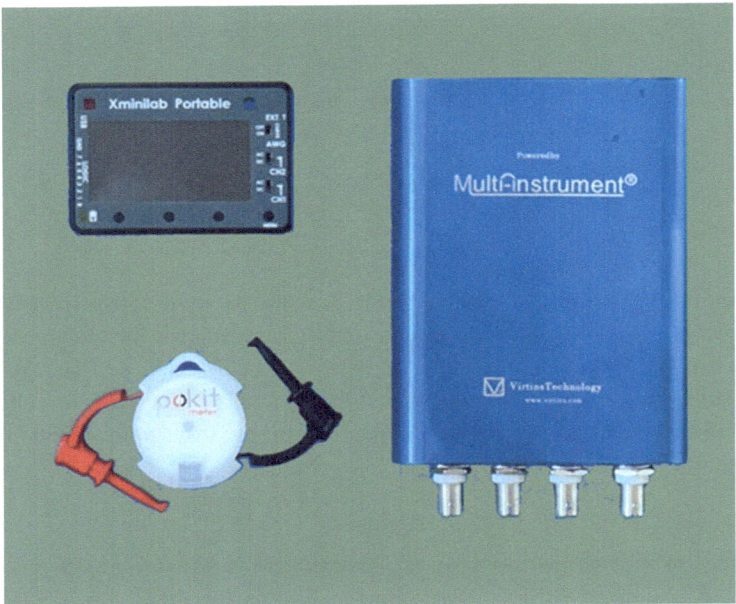

Fig. 71 Some inexpensive oscilloscopes: Top left—Handheld miniature, Bottom left—network connected using smart phone, Right—USB connected using laptop computer

Fig. 72 A picture of the beam with two different inexpensive sensors for measuring motion

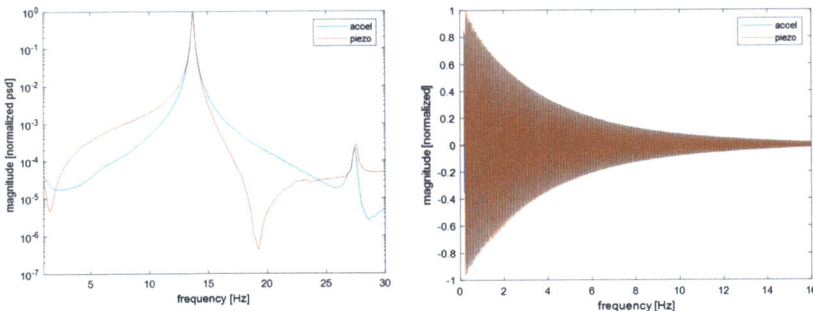

Fig. 73 The normalized power spectra (left) and time response (right) using an accelerometer and a piezoceramic sensor

References

1. **Nashif, Ahid, Jones, David and Henderson, John.** *Vibration Damping.* New York : Wiley, 1985.
2. **Inman, Daniel.** *Engineering Vibration.* Hoboken : Prentice Hall, 2014.
3. **Preumont, A.** *VIbration Control of Active Structures.* Berlin : Springer-Verlag, 2011.
4. **Frahm, Hermann.** *Device for damping vibrations of bodies. US989958A* United States of America, 04 18, 1911.
5. **Korenev, B. G.** *Dynamic VIbration Absorbers.* s.l. : Wiley, 1993.
6. *MD-80 Aft Cabin Noise Control.* **Lang, M. A., Lorch, D. R. and Simpson, M. A.** Friedrichschafen : NASA, 1992.
7. *Active vibration control using circular force generators.* **Black, Paul.** Munich : European Rotorcraft Forum, 2015.
8. **Moog.** Instructional link. [Online] https://www.youtube.com/watch?v=HDa1VO1VDpc.
9. **Griffin, Steven and Hovik, Stefan.** *Semi-active tuned mass damper to eliminate limit-cycle oscillation. US10539201B2* US, 2020.
10. **Wikipedia.** Aermet. [Online] https://en.wikipedia.org/wiki/Aermet.
11. **Moog.** TMD M-series data sheet. *CSA/Moog Web site.* [Online] csaengineering.com.
12. **Airpot.** [Online] www.airpot.com.
13. **Hovik, Stefan, et al.** *Variable spring-constant tuned mass damper. US10309867B2* 2019.
14. **Griffin, Steven and Aston, Richard, Martinez, Jazzmin, Langmack, Michael, Shurilla, Christopher.** *Systems and methods for dampening dynamic loading. US10364860* 2019.
15. **Piersol, Allan and Paez, Thomas.** *Harris' Shock and Vibration Handbook.* New York : McGraw Hill, 2010.
16. **Siemens.** NX Nastran User's Guide. [Online] https://docs.plm.automation.siemens.com/data_services/resources/nxnastran/10/help/en_US/tdocExt/pdf/User.pdf.
17. **Thomson, W. and Dahleh, M.** *Theory of Vibrations with Applications.* s.l. : Prentice-Hall, 1998.
18. **Motran.** Electromagnetic Linear Actuators For Active Vibration Control. [Online] http://www.motran.com/actuator-technology.html.

19. **Snowdon, J.** *Vibration and Shock in Damped Mechanical Systems.* New York : John Wiley & Sons, 1968.
20. *Vibration of damped plate oscillatory systems.* **Nicholas, J. and Bergman, L.** s.l. : Journal of Engineering Mechanics, 1986, Vol. 112.
21. *Three-dimensional architected materials and structures: Design, fabrication and mechanical behavior.* **Greer, J. and Deshpande, V.** 10, s.l. : MRS Bulletin, 2019, Vol. 44.
22. *Frequency Separation in Architected Structures Using Inverse Methods.* **Inman, D. and Gunasekar, A.** 4, s.l. : Journal of Applied and Computational Mechanics, 2021, Vol. 7.
23. **Griffin, S. and Niedermaier, D.** *Self-tuning tunable mass dampers. US9587699* 2017.
24. **Griffin, S. and Hovik, S.** *Semi-active tuned mass damper to eliminate limit-cycle oscillation. US10539201* 2020.
25. **Wikipedia.** Taipei 101. [Online] https://en.wikipedia.org/wiki/Taipei_101.
26. **Bramwell, A., Balmford, D. and Done, G.** *Bramwell's Helicopter Dynamics.* s.l. : Butterworth-Heinemann, 2001.
27. *Estimation of Modal Mass Using H infinity Optimal Model Reduction.* **Hwang, J. and Kim, H.** Seoul : CTBUH 2004 Seoul Conference, 2004.
28. **Griffin, Steven.** Repository for supplementary materials to Design and Test of Vibration Reducing Devices.
29. **3M.** *Technical Bulletin: Polymers Useful for Damping.* 2003.

SPRINGER NATURE

GPSR Compliance

The European Union's (EU) General Product Safety Regulation (GPSR) is a set of rules that requires consumer products to be safe and our obligations to ensure this.

If you have any concerns about our products, you can contact us on ProductSafety@springernature.com

In case Publisher is established outside the EU, the EU authorized representative is:

Springer Nature Customer Service Center GmbH
Europaplatz 3
69115 Heidelberg, Germany

The manufacturer's authorised representative in the EU is Springer Nature Customer Service Centre GmbH, Europaplatz 3, 69115 Heidelberg, Germany. If you have any concerns regarding our products, please contact ProductSafety@springernature.com

Printed and bound by CPI Group (UK) Ltd, Croydon, CR0 4YY

26/03/2026

02078977-0004